THE MULTIMESSENGER CHAKRA OF BLAZAR JETS

IAU SYMPOSIUM 375

COVER ILLUSTRATION:

Artistic impression of Hindu lord Vishnu holding an accretion disk with a jet. In Hindu sculptures in Nepal and India, the accretion disk pictorially resembles lord Vishnu's Sudarshan Chakra, conceived as a deity of destruction.

IAU SYMPOSIUM PROCEEDINGS SERIES

INTERNATIONAL ASTRONOMICAL UNION

UNION ASTRONOMIQUE INTERNATIONALE

International Astronomical Union

THE MULTIMESSENGER CHAKRA OF BLAZAR JETS

PROCEEDINGS OF THE 375th SYMPOSIUM OF THE INTERNATIONAL ASTRONOMICAL UNION KATHMANDU, NEPAL 5–9 DECEMBER, 2022

Edited by

IOANNIS LIODAKIS

Finnish Centre for Astronomy with ESO, Finland

MARGO F. ALLER

University of Michigan, USA

HENRIC KRAWCZYNSKI

Washington University St. Louis, USA

ANNE LÄHTEENMÄKI

Aalto University, Finland

and

TIMOTHY J. PEARSON

California Institute for Technology, USA

CAMBRIDGE UNIVERSITY PRESS

CAMBRIDGE UNIVERSITY PRESS
University Printing House, Cambridge CB2 8BS, United Kingdom
1 Liberty Plaza, Floor 20, New York, NY 10006, USA
10 Stamford Road, Oakleigh, Melbourne 3166, Australia

First published 2023

Printed in Great Britain by Henry Ling Limited, The Dorset Press, Dorchester, DT1 1HQ

Typeset in System LaTeX 2ε

A catalogue record for this book is available from the British Library Library of Congress Cataloguing in Publication data

This journal issue has been printed on FSC™-certified paper and cover board. FSC is an independent, non-governmental, not-for-profit organization established to promote the responsible management of the world's forests. Please see www.fsc.org for information.

ISBN 9781009353014 hardback
ISSN 1743-9213

Table of Contents

Preface

This special volume includes contributions from the IAU 375 symposium "The multimessenger chakra of blazar jets" that took place in Kathmandu Nepal during 5–9 December 2022 and included 91 participants from 25 countries across the world. The symposium lasted for 5 days covering all aspects of blazar science. The key science themes included multiwavelength/multimessenger theory and observations, particle acceleration mechanisms, jet structure, formation, composition and acceleration, multiwavelength polarization, as well as future experiments. The chapters of this volume have been organized to reflect the topics covered in each of the days.

Blazars are among the most intriguing and consistently bright objects in the observable Universe. They are the most extreme active galactic nuclei with powerful relativistic jets extending out to kpc from the central engine. Understanding how blazars form and shine has been a cumbersome endeavor since their discovery in the 1960s. Several fundamental questions regarding their intrinsic properties and the properties of the supermassive black holes in their centers are open to this day. 2020s mark the beginning of a new era, an era of large scale surveys, multimesseger astrophysics, high-energy polarization, and extreme angular resolution, setting the ideal stage to study astrophysical jets. The symposium had three goals. First, to bring together experts from all different aspects of the blazar community and facilitate the building of new collaborative efforts just in time to take advantage of the wealth of new incoming data that will help finally provide answers to long standing questions. Second, as the first IAU symposium in Nepal, to establish collaborative bridges between scientists from Nepal and scientists from around the World. Third, to support ongoing local efforts to promote astrophysics, and astrophysical research in Nepal.

To achieve a more interactive meeting and bridge the senior-junior scientist gap we organized mini workshops during lunch breaks on different topics related to the scientific theme of the day. We also organized "junior-senior lunches" by randomly assigning senior and junior scientists to different lunch groups giving the opportunity to break the ice between participants and allow junior scientist to gain the most out of the meeting.

Finally, throughout the symposium, we organized a vivid and diverse outreach program led by the Nepal Astronomical Society (NASO). The program included visits to schools by the symposium's participants, on-site training seminars on teaching astronomy for high-school teachers, and public talks. We also coordinated with RAD@home India and organized training sessions for citizen science related to radio galaxies and blazars.

Ioannis Liodakis
On behalf of the organizers

Editors

Ioannis Liodakis, Finnish Centre for Astronomy with ESO, Finland
Margo F. Aller, University of Michigan, USA
Henric Krawczynski, Washington University St. Louis, USA
Anne Lähteenmäki, Aalto University, Finland
Timothy J. Pearson, California Institute for Technology, USA

SOC Co-chairs

Roger Blandford, Stanford University, USA
Vasiliki Pavlidou, University of Crete, Greece

SOC members

Keiichi Asada, ASIAA, Taiwan
Geoffrey Bicknell, Australian National University, Australia
Carolina Casadio, Institute of Astrophysics, Greece
Annalisa Celotti, SISSA, Italy
Noemie Globus, University of California Santa Cruz, USA
Jose Luis Gomez, Instituto de Astrofisica de Andalucia , Spain
Talvikki Hovatta, University of Turku, Finland
Svetlana Jorstad, Boston University, USA
Yuri Kovalev, Max Planck Institute for Radioastronomy, Germany & Lebedev Physical Institute, Russia
Sang-Sung Lee, KASI, South Korea
Masanori Nakamura, ASIAA, Taiwan
Rodrigo Nemmen, University of Sao Paolo, Brazil
Maria Petropoulou, University of Athens, Greece
Claudia Raiteri, Osservatorio Astrofisico di Torino, Italy
Anamparambu Ramaprakash, IUCAA, India
Stas Shabala, University of Tasmania, Australia
Makoto Uemura, University of Hiroshima, Japan

LOC Co-chairs

Suresh Bhattarai, Nepal Astronomical Society, Nepal
Manisha Dwa, Nepal Astronomical Society, Nepal

LOC members

Dmitry Blinov, Institute of Astrophysics, Greece
Dinesh Kandel, Stanford University, USA
Ioannis Liodakis, Finnish Centre for Astronomy with ESO, Finland

List of Participants

Agniva Roychowdhury, University of Maryland, Baltimore County
Aleksandr Popkov, Moscow Institute of Physics and Technology
Alessandro Paggi, University of Turin
Alexander Plavin, Lebedev Physical Institute
Alice Pasetto, ITyA-UNAM
Amal Abdulrahman, Department of Physics, Farook College, Calicut University
Amar Raj Ghimire, Tribhuvan University
Aminabi Thekkoth, University of Calicut
Amir Levinson, Tel Aviv University
Anabella Araudo, LUPM (CNRS) and ELI Beamlines (Czech Academy of Sciences)
Anamparambu Ramaprakash, IUCAA
Ananda Hota, UM-DAE CEBS & RAD@home, India
Anastasiia Omeliukh, Ruhr University Bochum
Andrzej Zdziarski, N. Copernicus Astronomical Center
Annalisa Celotti, SISSA, Italy
Anthony Readhead, California Institut of Technology
Athira M Bharathan, CHRIST Deemed to be University
Avinash Kumar, RAD@home India
Baheeja Cholakkal, University of Calicut
Callum McCall, Liverpool John Moores University
Carolina Casadio, Institute of Astrophysics - FORTH
David Fernandez, Sejong University
Dmitry Blinov, IA, FORTH
Eileen Meyer, University of Maryland Baltimore County
Elena Nokhrina, Moscow Institute of Physics and Technology
Elena Shablovinskaya, Special astrophysical observatory RAS
Eli Kasai, University of Namibia
Elina Lindfors, Finnish Centre for Astronomy with ESO, University of Turku
Erin O'Sullivan, Uppsala University
Evaristus Iyida, Department of Physics and Astronomy, University of Nigeria, Nsukka
Florian Eppel, JMU Wuerzburg
Florian Roesch, JMU Wuerzburg
Francesco Massaro, University of Turin
Gabriele Giovannini, IRA/INAF and Bologna University
Georgios Filippos Paraschos, Max Planck Institute for Radio Astronomy
Gourab Giri, Indian Institute of Technology Indore
Haocheng Zhang, NPP Fellow/NASA GSFC
Ioannis Liodakis, Finnish Centre for Astronomy with ESO, University of Turku
Ishika Palit, Tel Aviv university, Israel
Ivan Agudo, IAA-CSIC
Jae-Young Kim, Kyungpook National University (KNU)
Janhavi Baghel, National Centre for Radio Astrophysics, Pune, India
Jeffrey Hodgson, Sejong University
Joana Kramer, Max Planck Institute for Radio Astronomy
Jun Yi (Kevin) Koay, Academia Sinica Institute of Astronomy and Astrophysics
Kenichi Nishikawa, Alabama A&M University
Kenji Yoshida, Shibaura Institute of Technology
Konstantinos Tassis, IA-FORTH and University of Crete
Koushik Chatterjee, Harvard University
Krzysztof Nalewajko, Nicolaus Copernicus Astronomical Center, Polish Academy of Sciences

Lawrence Peirson, Stanford University
Lea Heckmann, Max Planck Institute for Physics
Lea Marcotulli, Yale University
Leonid Gurvits, Joint Institute for VLBI ERIC and Delft University of Technology
Lorena Hernandez-Garcia, MAS/UV (Chile)
Luca Ighina, Uni. Insubria + INAF-Brera
Maria Charisi, Vanderbilt University
Markus Boettcher, North-West University
Megha Rajoria, RAD@home India
Monika Moscibrodzka, Radboud University
Naoki Isobe, ISAS/JAXA
Nicholas MacDonald, Max Planck Institute for Radio Astronomy
Olivier Hervet, UC Santa Cruz
Po-Chih Hsu, Graduate Institute of Astronomy, National Central University, Taiwan
Preeti Kharb, National Centre for Radio Astrophysics - Tata Institute of Fundamental Research
Rajesh Kumar Bachchan, Tribhuvan University
Ravi Pratap Dubey, Max Planck Institute for Astronomy
Ryo Imazawa, Hiroshima University
Sagar Chapagain, Tribhuvan University
Sasha tcheskovskoy, Northwestern University
Shivangi Pandey, Aryabhatta Research Institute of Observational Sciences, Nainital
Soeb Razzaque, University of Johannesburg
Sriyasriti Acharya, Indian Institute of Technology Indore
Sushma Bashyal, Tribhuvan University
Susmita Das, Presidency University, Kolkata, India
Susumu Inoue, Bunkyo U. / RIKEN
Swaraj Pradhan, Nepal Astronomical Society, Astronomers Without Boarders
Tapio Pursimo, Nordic Optical Telescope / Aarhus Universitet
Tej Bahadur Chand, Centre for Space Research, North-West University Potchefstroom, South Africa
Thalia Traianou, IAA-CSIC
Tullia Sbarrato, INAF - Osservatorio Astronomico di Brera
Tuomas Savolainen, Aalto University
Uwe Bach, Max-Planck-Institut fuer Radioastronomie
Vasiliki Pavlidou, IA-FORTH and University of Crete
Vasily Beskin, Lebedev Physical Institute & Moscow Institute of Physics and Technology
Yamuna Rana, Tribhuvan University
Yogesh Singh Maharjan, Tribhuvan University
Yosuke Mizuno, Tsung-Dao Lee Institute, Shanghai Jiao Tong University
Yuh Tsunetoe, Kyoto University
Yuri Kovalev, MPIfR
Zahoor Ahmad Malik, University of Kashmir

Group photograph of the Symposium's participants at the venue.

Group photograph of the Symposium's participants during the excursion at the UNESCO's world heritage site of Bhaktapur.

The Multimessenger Chakra of Blazar Jets
Proceedings IAU Symposium No. 375, 2023
I. Liodakis, M. F. Aller, H. Krawczynski, A. Lähteenmäki & T. J. Pearson, eds.
doi:10.1017/S1743921323000911

Extragalactic Jets from Radio to Gamma-rays

E. Meyer[1], **A. Shaik**[1], **K. Reddy**[2] **and M. Georganopoulos**[1,3]

[1]University of Maryland Baltimore County, 1000 Hilltop Cir, Baltimore MD 21250, USA

[2]School of Earth and Space Exploration, Arizona State University, Tempe, AZ 85287, USA

[3]NASA Goddard Space Flight Center, Greenbelt, MD USA

Abstract. Despite the fact that jets from black holes were first understood to exist over 40 years ago, we are still in ignorance about many primary aspects of these systems – including the radiation mechanism at high energies, the particle makeup of the jets, and how particles are accelerated, possibly to energies as high as 100 TeV and hundreds of kpc from the central engine. We focus in particular on the discovery (and mystery) of strong X-ray emission from radio jets on kpc-scales, enabled by the unequaled high resolution of the *Chandra* X-ray observatory. We review the main evidence for and against the viable models to explain this X-ray emission over the last 20 years. Finally, we present results of a recent study on the X-ray variability of kpc-scale jets, where we find evidence that between 30-100% of the X-ray jet population is variable at the tens-of-percent level. The short (~years) variability timescale is incompatible with the IC/CMB model for the X-rays and implies extremely small structures embedded within the kpc-scale jet, and thus requires a reconsideration of many assumptions about jet structure and dynamics.

Keywords. AGN, Blazar, Jets, Variability

1. Introduction

Active galactic nuclei (AGN) are powered by the supermassive black holes (SMBH) which reside in the centers of all relatively massive galaxies (Kormendy and Ho 2013). A small percentage of these sources eject bipolar, collimated jets of relativistic plasma which emit brightly at radio and sometimes optical and X-ray frequencies kiloparsecs (kpc) away from the central engine. Resolved radio jets are often separated into two classes based on radio morphology: Fanaroff & Riley (1974) class I (FRI) jets have plume-like jets and are dominated by emission near the core, while FRII jets are highly collimated and are dominated by bright hotspots where the jet impacts into the intergalactic medium (IGM). Historically, this classification system was also associated with a difference in jet power, with FRI and FRII jets representing low- and high-power jets respectively. However, more recent studies have discovered low-power FRII galaxies (Mingo et al. 2019) and large population studies suggest a large range in jet power overlap (Keenan et al. 2021).

One of the major discoveries by the *Chandra* X-ray Observatory has been the detection of X-rays from radio jets on kpc-scales (Chartas et al. 2000; Schwartz et al. 2000; Harris and Krawczynski 2006; Worrall 2009; Marshall et al. 2018). Assuming a leptonic jet model, the X-ray emission of most low-power or FRI class sources, including nearby sources such as M87 (Harris 2003) and Cen A (Hardcastle et al. 2007), are often (but not always) well-described by synchrotron radiation from a single electron population which extends from radio to optical and X-ray energies. However in many (typically the most powerful) quasar-hosted jets, the X-ray emission is too high, and the spectral index too

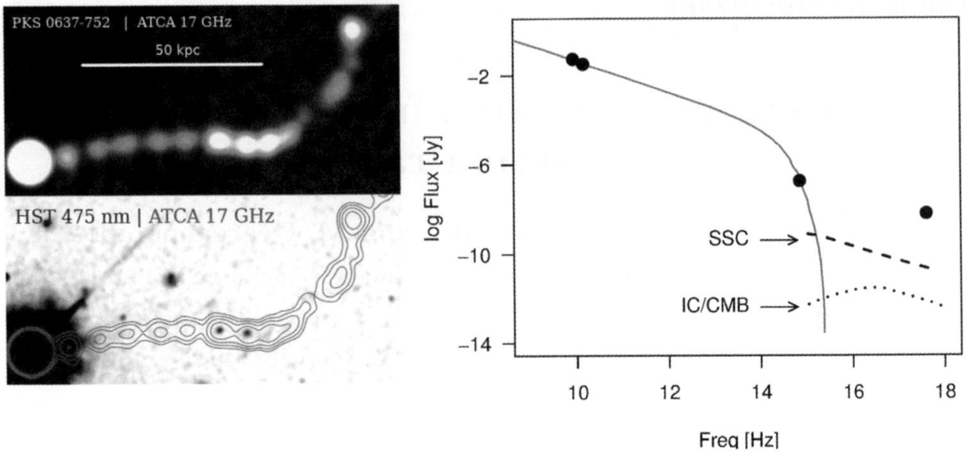

Figure 1. Upper left, an ATCA radio image of PKS 0637-752 (provided by Leith Godfrey), below, an HST image of the jet with inverted color scale and radio contours overlaid. Note the small knots of optical emission coinciding with the brightest radio knots. At right, an early SED of the bright knots of the jet, showing the very high level of X-ray emission and the difficulty of explaining it with either SSC or (unbeamed) IC/CMB.

hard, to be attributed to the low-energy synchrotron component. A classic example (and the first X-ray source *Chandra* observed) of PKS 0637-752 is shown in Figure 1. The observation of 'anomalously' bright and hard X-rays has since been extended to virtually all powerful FRII jets and even some low-power FRI jets such as in the case of M84 (Meyer et al. 2018).

2. The Rise and Fall of the IC/CMB model

The most commonly adopted explanation for the 'anomalous' X-ray emission from kpc-scale jets is inverse-Compton scattering of Cosmic Microwave Background (IC/CMB) photons by a still-relativistic jet. Under this model, high-energy electrons in the jet upscatter low-energy CMB photons, provided that the jet is both highly relativistic with high bulk Lorentz factors ($\Gamma > 10$) on kpc-scales and closely aligned to our line-of-sight ($\theta < 5°$; Tavecchio et al. 2000; Celotti et al. 2001). This produces the high Doppler boosting required to reproduce the observed bright and hard-spectrum X-ray flux. However, this emission mechanism also sometimes predicts kinetic powers in excess of the Eddington limit (Atoyan and Dermer 2004) and is in many cases at odds with other more recent observations, including significantly higher polarization of the ultraviolet/X-ray component than expected (Cara et al. 2013) and jet-counterjet flux ratios which are inconsistent with IC/CMB predictions (Kataoka et al. 2008; Clautice et al. 2016; Hardcastle et al. 2016).

Another major effort at testing the IC/CMB model has been through looking for the high levels of GeV emission that it predicts, first proposed before the launch of *Fermi* by Georganopoulos et al. (2006). The analysis is complicated by the poor (fractions of a degree) resolution of *Fermi* as it is impossible to spatially distingues the jet from the core, which is highly variable and is expected dominate over the jet especially at lower (MeV) energies. However, through the use of a recombined light-curve analysis, Meyer and Georganopoulos (2014) first showed that the IC/CMB model could be strongly ruled out through gamma-ray upper limits for the X-ray jet of 3C 273, before doing the same in PKS 0637-752 (Meyer et al. 2015, 2017). Peter Breiding then widely applied this test to essentially all the 'multiple spectral component' or MSC jets for his PhD

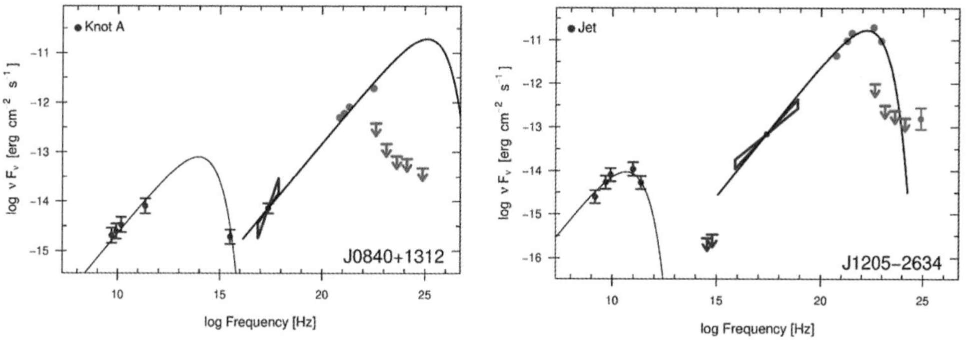

Figure 2. Example large-scale (kpc-scale) jet SEDs, with synchrotron (thin line) and IC/CMB (thick line) model curves. As shown, matching the X-ray emission unavoidably predicts a high level of GeV emission, which is in violation with the deep *Fermi*/LAT upper limits (red). Taken from Breiding et al. (2023).

thesis work (Breiding et al. 2017, 2023), finding that 24/45 sources tested would require IC/CMB-predicted GeV emission significantly in violation of the gamma-ray limits (see example SEDs for two of the sources from Breiding et al. (2023) in Figure 2).

Another difficulty for the IC/CMB model is the change in apparent morphology (knot location, size, etc) from radio to optical to X-ray. The simplest IC/CMB model requires cospatial emission in these different bands, since the radio-to-optical synchrotron spectrum is produced by the same electron energy distribution upscattering the CMB to X-ray energies (or if there is an offset, for the X-rays to persist beyond the radio since these arise from very low-energy electrons in the IC/CMB interpretation). However, an extensive study of essentially all X-ray jets discovered to date finds significant offsets between X-rays and radio on the order of ∼1 kpc or more (e.g., Reddy et al. 2021, 2022).

3. Variability of X-ray Jets

One consequence of the IC/CMB model is a low energy extension of the electron energy distribution. That is, in order for IC/CMB to be the dominant X-ray emission mechanism, there must be a significant population of particles at low-energies ($\gamma_{min} \sim 10-20$; Celotti et al. 2001; Georganopoulos et al. 2006). Besides the considerable energy requirements of this extension, the implied radiative cooling times are extremely long (longer than the jet lifetimes), which means that under an IC/CMB X-ray mechanism we should absolutely not expect to see variability in X-ray emission on timescales within the lifetime of *Chandra*.

It was thus surprising when Marshall et al. (2010) examined the case of the nearby ($z = 0.035$) FRII Pictor A, a radio galaxy with a very long (arcminute-scale) radio jet and steep radio core, and a clearly visible counterjet. A knot in the jet of Pictor A was seen to fade over a timescale of a few years with a reported significance of 3.4σ. Hardcastle et al. (2016) and Thimmappa et al. (2020) later confirmed these results at similar significances ($p < 0.011$). While variability had been observed by this point in nearby FRI jets (most dramatically in the mid-2000s outburst of HST-1 in the jet of M87; Harris et al. 2003) variability on this timescale was completely unexpected. In the case of Pictor A, assuming minimum energy conditions and an emitting region the size of the jet cross-section, the synchrotron cooling timescale is on the order of 1200 years, implying that the X-rays arise from a region much smaller and with a magnetic field several times higher than equipartition.

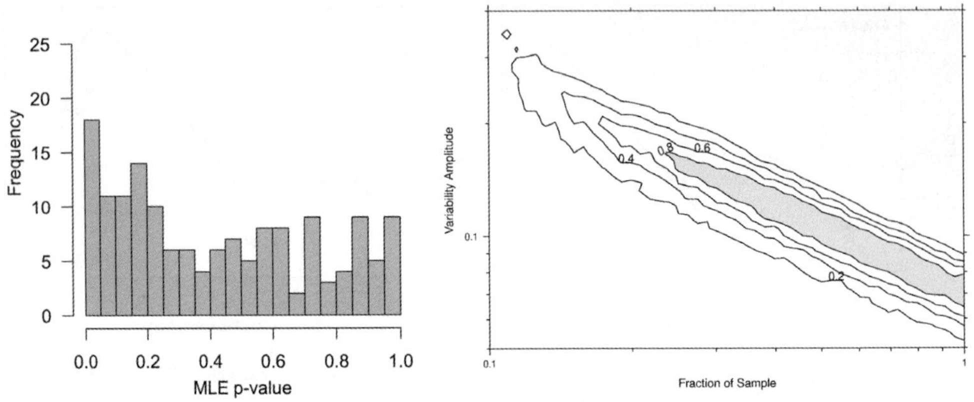

Figure 3. At left, the histogram of p-values from our likelihood model analysis of 155 jet regions from 53 total sources. Under the null hypothesis of steady emission, we expect a completely uniform (0,1) distribution of p-values. The global p-value associated with the excess of low values shown here is p=0.00019, suggesting variability in the jet population. At right, the results of a simulation to test the degeneracy between sample fraction of variability (assuming 1-f are steady emitters) and the typical variability amplitude (relative to mean). The shaded blue region shows the best agreement with the observational data, suggesting between 30-100% of jets are variable, at a modest tens-of-percent scale.

3.1. *New Results*

Over the past few years, we have conducted a comprehensive archival study of X-ray jets, to look for variability in other sources. Our sample of 53 comprises nearly all known X-ray jets imaged more than once by the *Chandra* Advanced CCD Imaging Spectrometer (ACIS) instrument. The average number of observations per source in our sample is 3.4, with a mean spacing of 2.6 years. Two sources (3C 305 and 3C 171) were excluded as the extended X-ray emission associated has been attributed to jet-driven gas (Hardcastle et al. 2012, 2010); we also exclude the sources M87 and Centaurus A, which are already known to produce X-rays by synchrotron process (Feigelson et al. 1981; Kraft et al. 2002; Wilson et al. 2000). Both M87 and Centaurus A have been far more deeply and frequently observed than typical of the remaining sample (with 50 and 43 distinct observations, respectively), and this is likely one of the reasons that X-ray variability has been reported in both cases. With the exception of Pictor A, variability has not been reported for any other source in our sample of 53. Unlike the jets of M87 and Centaurus A, however, the X-ray emission in Pictor A is clearly from a second emission component (Hardcastle and Croston 2005).

Our likelihood function is a straightforward application of Poisson statistics, with a null hypothesis of a steady source rate for each individual knot region, taking into account a varying background and various detector effects. We computed for each knot a p-value for the test of the the null hypothesis of a steady source rate, the distribution of which (for 155 test regions) is shown in Figure 3. Out of the full sample of 155 regions tested, 18 (12%) have p-values less than 0.05, suggesting significant variability in the intrinsic source rate. The single-region p-values for the are expected to follow a uniform $U(0,1)$ distribution under the null hypothesis of steady emission, so this is clearly about twice the value expected, and the excess of low p-values is apparent by eye in Figure 3. This is also born out in the statistics: when we compare all 155 single-region p-values to a $U(0,1)$ distribution using a one-sided Kolmogorov-Smirnov (KS) test, we obtain a global p-value of 0.000196, indicating that the distribution is highly non-Uniform. This clearly indicates that the observations are not consistent with non-variable X-ray sources.

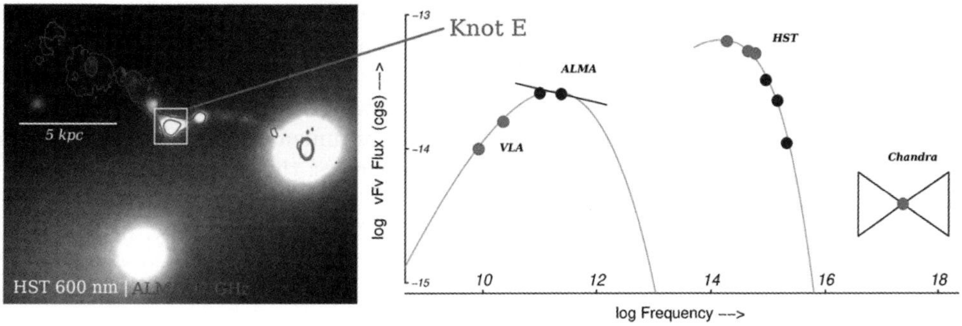

Figure 4. At left, an HST image of the jet of 3C 346, with ALMA band3 contours overlaid. At right, a detailed SED of knot E, showing at least 3 emission components in the jet, so far unexplained (Meyer et al., in prep.)

Figure 3 (right panel) shows the results of a simulation to try to constrain the typical scale of variability in our jet population, given a degree of degeneracy between the typical variability amplitude and rate of variable sources existing in the population. It is not implausible that a subset of X-ray jets are steady X-ray emitters while the rest exhibit variability. For example, 30% of the sample might be variable sources, with a characteristic scale of 50% in amplitude, or 90% of the sample might be variable with a lower (say, 10%) characteristic amplitude of variability, and both of these scenarios might be consistent with the *p*-value distribution we observe. Our simulation compares the MLE *p*-value distribution of simulated populations varying in these two respects to our actual *p*-value distribution and reports the K-S test *p*-value for the comparison. Thus, the lighter-colored regions of the plot are most consistent with the data while black areas are not consistent with the data under a K-S test comparison.

4. Discussion

As a result of this and the other observational lines of evidence, the *general* validity of IC/CMB model for large-scale jet X-ray emission must be called into question. On the other hand, IC/CMB does appear to be more consistent than alternatives in some cases, for example the few high-redshift jets with unusually faint or undetected radio emission (e.g. Simionescu et al. 2016; Migliori et al. 2022), and potentially in somewhat 'unusual' cases at lower redshift. For example, we found that the 'plateau' or underlying steady-state gamma-ray emission in two X-ray jets (PKS 1510-089 and OJ 287) was compatible with the predicted IC/CMB level (Meyer et al. 2018). However, these cases are likely outliers, with a particularly favorable alignment and unusually high jet speed at large distances.

The main alternative to the IC/CMB model is synchrotron radiation from a second electron energy distribution (EED), or abandoning the leptonic model in favor of a hadronic jet model (Petropoulou et al. 2017; Meyer et al. 2018). Synchrotron models allow us to avoid the 'uncomfortable' requirements of IC/CMB models (e.g., high bulk Lorentz factors, small viewing angle, and super-Eddington jet powers), but still have a largely ad-hoc appearance: multiple EEDs was not predicted by theory. Indeed, an example of a kpc-scale jet region with three distinct emission components is shown in Figure 4; here the optical and X-ray emission appear to arise from distinct and unknown processes. Previously, the apparent co-spatiality of the radio and X-rays seemed consistent with a single population (i.e. IC/CMB) X-ray model, but recent research has shown that they are in fact *not* co-spatial in most cases (Reddy et al. 2022). Notably, these different mechanisms along with IC/CMB involve vastly different scenarios as far as jet matter content and energy distribution, jet power, and acceleration mechanisms, with important implications for a proper accounting of jet impacts on the environment.

The short timescales of X-ray variability observed in our study implies that the emitting regions are much smaller that the width of the jet (which is generally resolved in e.g., radio to be on kpc scales). By the light crossing time argument a flare event on the order of a year cannot occur in a region larger than a few parsecs; this requires highly localized particle acceleration and appears more consistent with magnetic reconnection than the usually assumed shock acceleration (Giannios et al. 2009). Interestingly, production of high energy photons from regions much smaller than the source physical size has been observed in very different celestial objects, from the Sun (Omodei et al. 2013) and the Crab Nebula (Abdo et al. 2011), to blazars like PKS 1510-089, where variations over hours are seen in the VLBI radio core that has a light crossing time of the order of a year, a difference too large to explain through beaming (Marscher and Jorstad 2010). These observations are key to understanding the particle acceleration mechanism acting in these environments.

References

Abdo, A. A., Ackermann, M., Ajello, M., Allafort, A., Baldini, L., Ballet, J., Barbiellini, G., Bastieri, D., Bechtol, K., Bellazzini, R., Berenji, B., Blandford, R. D., Bloom, E. D., Bonamente, E., Borgland, A. W., Bouvier, A., Brandt, T. J., Bregeon, J., Brez, A., Brigida, M., Bruel, P., Buehler, R., Buson, S., Caliandro, G. A., Cameron, R. A., Cannon, A., Caraveo, P. A., Casandjian, J. M., Çelik, Ö., Charles, E., Chekhtman, A., Cheung, C. C., Chiang, J., Ciprini, S., Claus, R., Cohen-Tanugi, J., Costamante, L., Cutini, S., D'Ammando, F., Dermer, C. D., de Angelis, A., de Luca, A., de Palma, F., Digel, S. W., do Couto e Silva, E., Drell, P. S., Drlica-Wagner, A., Dubois, R., Dumora, D., Favuzzi, C., Fegan, S. J., Ferrara, E. C., Focke, W. B., Fortin, P., Frailis, M., Fukazawa, Y., Funk, S., Fusco, P., Gargano, F., Gasparrini, D., Gehrels, N., Germani, S., Giglietto, N., Giordano, F., Giroletti, M., Glanzman, T., Godfrey, G., Grenier, I. A., Grondin, M. H., Grove, J. E., Guiriec, S., Hadasch, D., Hanabata, Y., Harding, A. K., Hayashi, K., Hayashida, M., Hays, E., Horan, D., Itoh, R., Jóhannesson, G., Johnson, A. S., Johnson, T. J., Khangulyan, D., Kamae, T., Katagiri, H., Kataoka, J., Kerr, M., Knödlseder, J., Kuss, M., Lande, J., Latronico, L., Lee, S. H., Lemoine-Goumard, M., Longo, F., Loparco, F., Lubrano, P., Madejski, G. M., Makeev, A., Marelli, M., Mazziotta, M. N., McEnery, J. E., Michelson, P. F., Mitthumsiri, W., Mizuno, T., Moiseev, A. A., Monte, C., Monzani, M. E., Morselli, A., Moskalenko, I. V., Murgia, S., Nakamori, T., Naumann-Godo, M., Nolan, P. L., Norris, J. P., Nuss, E., Ohsugi, T., Okumura, A., Omodei, N., Ormes, J. F., Ozaki, M., Paneque, D., Parent, D., Pelassa, V., Pepe, M., Pesce-Rollins, M., Pierbattista, M., Piron, F., Porter, T. A., Rainò, S., Rando, R., Ray, P. S., Razzano, M., Reimer, A., Reimer, O., Reposeur, T., Ritz, S., Romani, R. W., Sadrozinski, H. F. W., Sanchez, D., Parkinson, P. M. S., Scargle, J. D., Schalk, T. L., Sgrò, C., Siskind, E. J., Smith, P. D., Spandre, G., Spinelli, P., Strickman, M. S., Suson, D. J., Takahashi, H., Takahashi, T., Tanaka, T., Thayer, J. B., Thompson, D. J., Tibaldo, L., Torres, D. F., Tosti, G., Tramacere, A., Troja, E., Uchiyama, Y., Vandenbroucke, J., Vasileiou, V., Vianello, G., Vitale, V., Wang, P., Wood, K. S., Yang, Z., & Ziegler, M. 2011, Gamma-Ray Flares from the Crab Nebula. *Science*, 331(6018), 739.

Atoyan, A. & Dermer, C. D. 2004, Synchrotron versus Compton Interpretations for Extended X-Ray Jets. *ApJ*, 613(1), 151–158.

Breiding, P., Meyer, E. T., Georganopoulos, M., Keenan, M. E., DeNigris, N. S., & Hewitt, J. 2017, Fermi Non-detections of Four X-Ray Jet Sources and Implications for the IC/CMB Mechanism. *ApJ*, 849(2), 95.

Breiding, P., Meyer, E. T., Georganopoulos, M., Reddy, K., Kollmann, K. E., & Roychowdhury, A. 2023, A multiwavelength study of multiple spectral component jets in AGN: testing the IC/CMB model for the large-scale-jet X-ray emission. *MNRAS*, 518(3), 3222–3250.

Cara, M., Perlman, E. S., Uchiyama, Y., Cheung, C. C., Coppi, P. S., Georganopoulos, M., Worrall, D. M., Birkinshaw, M., Sparks, W. B., Marshall, H. L., Stawarz, L., Begelman, M. C., O'Dea, C. P., & Baum, S. A. 2013, Polarimetry and the High-energy Emission Mechanisms in Quasar Jets: The Case of PKS 1136-135. *ApJ*, 773, 186.

Celotti, A., Ghisellini, G., & Chiaberge, M. 2001, Large-scale jets in active galactic nuclei: multiwavelength mapping. *MNRAS*, 321(1), L1–L5.

Chartas, G., Worrall, D. M., Birkinshaw, M., Cresitello-Dittmar, M., Cui, W., Ghosh, K. K., Harris, D. E., Hooper, E. J., Jauncey, D. L., Kim, D. W., Lovell, J., Mathur, S., Schwartz, D. A., Tingay, S. J., Virani, S. N., & Wilkes, B. J. 2000, The Chandra X-Ray Observatory Resolves the X-Ray Morphology and Spectra of a Jet in PKS 0637-752. *ApJ*, 542(2), 655–666.

Clautice, D., Perlman, E. S., Georganopoulos, M., Lister, M. L., Tombesi, F., Cara, M., Marshall, H. L., Hogan, B., & Kazanas, D. 2016, The Spectacular Radio-near-IR-X-Ray Jet of 3C 111: The X-Ray Emission Mechanism and Jet Kinematics. *ApJ*, 826, 109.

Feigelson, E. D., Schreier, E. J., Delvaille, J. P., Giacconi, R., Grindlay, J. E., & Lightman, A. P. 1981, The X-ray structure of Centaurus A. *ApJ*, 251, 31–51.

Georganopoulos, M., Perlman, E. S., Kazanas, D., & McEnery, J. 2006, Quasar X-Ray Jets: Gamma-Ray Diagnostics of the Synchrotron and Inverse Compton Hypotheses: The Case of 3C 273. *ApJ Letters*, 653(1), L5–L8.

Giannios, D., Uzdensky, D. A., & Begelman, M. C. 2009, Fast TeV variability in blazars: jets in a jet. *MNRAS*, 395(1), L29–L33.

Hardcastle, M. J. & Croston, J. H. 2005, The Chandra view of extended X-ray emission from Pictor A. *MNRAS*, 363(2), 649–660.

Hardcastle, M. J., Kraft, R. P., Sivakoff, G. R., Goodger, J. L., Croston, J. H., Jordán, A., Evans, D. A., Worrall, D. M., Birkinshaw, M., Raychaudhury, S., Brassington, N. J., Forman, W. R., Harris, W. E., Jones, C., Juett, A. M., Murray, S. S., Nulsen, P. E. J., Sarazin, C. L., & Woodley, K. A. 2007, New Results on Particle Acceleration in the Centaurus A Jet and Counterjet from a Deep Chandra Observation. *ApJ Letters*, 670(2), L81–L84.

Hardcastle, M. J., Lenc, E., Birkinshaw, M., Croston, J. H., Goodger, J. L., Marshall, H. L., Perlman, E. S., Siemiginowska, A., Stawarz, Ł., & Worrall, D. M. 2016, Deep Chandra observations of Pictor A. *MNRAS*, 455(4), 3526–3545.

Hardcastle, M. J., Massaro, F., & Harris, D. E. 2010, X-ray emission from the extended emission-line region of the powerful radio galaxy 3C171. *MNRAS*, 401(4), 2697–2705.

Hardcastle, M. J., Massaro, F., Harris, D. E., Baum, S. A., Bianchi, S., Chiaberge, M., Morganti, R., O'Dea, C. P., & Siemiginowska, A. 2012, The nature of the jet-driven outflow in the radio galaxy 3C 305. *MNRAS*, 424(3), 1774–1789.

Harris, D. E. 2003, X-ray variability and emission process of the radio jet in M87. *NewAR*, 47(6-7), 617–620.

Harris, D. E., Biretta, J. A., Junor, W., Perlman, E. S., Sparks, W. B., & Wilson, A. S. 2003, Flaring X-Ray Emission from HST-1, a Knot in the M87 Jet. *ApJ Letters*, 586(1), L41–L44.

Harris, D. E. & Krawczynski, H. 2006, X-Ray Emission from Extragalactic Jets. *ARA&A*, 44(1), 463–506.

Kataoka, J., Stawarz, Ł., Harris, D. E., Siemiginowska, A., Ostrowski, M., Swain, M. R., Hardcastle, M. J., Goodger, J. L., Iwasawa, K., & Edwards, P. G. 2008, Chandra Reveals Twin X-Ray Jets in the Powerful FR II Radio Galaxy 3C 353. *ApJ*, 685(2), 839–857.

Keenan, M., Meyer, E. T., Georganopoulos, M., Reddy, K., & French, O. J. 2021, The relativistic jet dichotomy and the end of the blazar sequence. *MNRAS*, 505(4), 4726–4745.

Kormendy, J. & Ho, L. C. 2013, Coevolution (Or Not) of Supermassive Black Holes and Host Galaxies. *ARA&A*, 51(1), 511–653.

Kraft, R. P., Forman, W. R., Jones, C., Murray, S. S., Hardcastle, M. J., & Worrall, D. M. 2002, Chandra Observations of the X-Ray Jet in Centaurus A. *ApJ*, 569(1), 54–71.

Marscher, A. P. & Jorstad, S. G. 2010, Rapid Variability of Gamma-ray Emission from Sites near the 43 GHz Cores of Blazar Jets. *arXiv e-prints*,, arXiv:1005.5551.

Marshall, H. L., Gelbord, J. M., Worrall, D. M., Birkinshaw, M., Schwartz, D. A., Jauncey, D. L., Griffiths, G., Murphy, D. W., Lovell, J. E. J., Perlman, E. S., & Godfrey, L. 2018, An X-Ray Imaging Survey of Quasar Jets: The Complete Survey. *ApJ*, 856(1), 66.

Marshall, H. L., Hardcastle, M. J., Birkinshaw, M., Croston, J., Evans, D., Landt, H., Lenc, E., Massaro, F., Perlman, E. S., Schwartz, D. A., Siemiginowska, A., Stawarz, Ł., Urry, C. M., & Worrall, D. M. 2010, A Flare in the Jet of Pictor A. *ApJ Letters*, 714(2), L213–L216.

Meyer, E. T., Breiding, P., Georganopoulos, M., Oteo, I., Zwaan, M. A., Laing, R., Godfrey, L., & Ivison, R. J. 2017, New ALMA and Fermi/LAT Observations of the Large-scale Jet of PKS 0637–752 Strengthen the Case Against the IC/CMB Model. *ApJ Letters*, 835, L35.

Meyer, E. T. & Georganopoulos, M. 2014, Fermi Rules Out the Inverse Compton/CMB Model for the Large-scale Jet X-Ray Emission of 3C 273. *ApJ Letters*, 780, L27.

Meyer, E. T., Georganopoulos, M., Sparks, W. B., Godfrey, L., Lovell, J. E. J., & Perlman, E. 2015, Ruling out IC/CMB X-rays in PKS 0637-752 and the Implications for TeV Emission from Large-scale Quasar Jets. *ApJ*, 805, 154.

Meyer, E. T., Petropoulou, M., Georganopoulos, M., Chiaberge, M., Breiding, P., & Sparks, W. B. 2018, Detection of an Optical/UV Jet/Counterjet and Multiple Spectral Components in M84. *ApJ*, 860(1), 9.

Migliori, G., Siemiginowska, A., Cheung, C. C., Celotti, A., Giroletti, M., Giovannini, G., Paggi, A., & Liuzzo, E. 2022, Discovery of a bright extended X-ray jet in RGB J1512+020A. *MNRAS*, 512(3), 4639–4659.

Mingo, B., Croston, J. H., Hardcastle, M. J., Best, P. N., Duncan, K. J., Morganti, R., Rottgering, H. J. A., Sabater, J., Shimwell, T. W., Williams, W. L., Brienza, M., Gurkan, G., Mahatma, V. H., Morabito, L. K., Prandoni, I., Bondi, M., Ineson, J., & Mooney, S. 2019, Revisiting the Fanaroff-Riley dichotomy and radio-galaxy morphology with the LOFAR Two-Metre Sky Survey (LoTSS). *MNRAS*, 488(2), 2701–2721.

Omodei, N., Petrosian, V., Pesce-Rollins, M., & the Fermi-LAT Collaboration 2013, Fermi-LAT Observation of Impulsive Solar Flares. *arXiv e-prints*,, arXiv:1304.0798.

Petropoulou, M., Vasilopoulos, G., & Giannios, D. 2017, The TeV emission of Ap Librae: a hadronic interpretation and prospects for CTA. *MNRAS*, 464(2), 2213–2222.

Reddy, K., Georganopoulos, M., & Meyer, E. T. 2021, X-Ray-to-radio Offset Inference from Low-count X-Ray Jets. *ApJS*, 253(2), 37.

Reddy, K., Georganopoulos, M., & Meyer, E. T. 2022, ATLAS-X. *ApJS*,.

Schwartz, D. A., Marshall, H. L., Lovell, J. E. J., Piner, B. G., Tingay, S. J., Birkinshaw, M., Chartas, G., Elvis, M., Feigelson, E. D., Ghosh, K. K., Harris, D. E., Hirabayashi, H., Hooper, E. J., Jauncey, D. L., Lanzetta, K. M., Mathur, S., Preston, R. A., Tucker, W. H., Virani, S., Wilkes, B., & Worrall, D. M. 2000, Chandra Discovery of a 100 kiloparsec X-Ray Jet in PKS 0637-752. *ApJ Letters*, 540(2), 69–72.

Simionescu, A., Stawarz, Ł., Ichinohe, Y., Cheung, C. C., Jamrozy, M., Siemiginowska, A., Hagino, K., Gandhi, P., & Werner, N. 2016, Serendipitous Discovery of an Extended X-Ray Jet without a Radio Counterpart in a High-redshift Quasar. *ApJ Letters*, 816(1), L15.

Tavecchio, F., Maraschi, L., Sambruna, R. M., & Urry, C. M. 2000, The X-Ray Jet of PKS 0637-752: Inverse Compton Radiation from the Cosmic Microwave Background? *ApJ Letters*, 544(1), L23–L26.

Thimmappa, R., Stawarz, Ł., Marchenko, V., Balasubramaniam, K., Cheung, C. C., & Siemiginowska, A. 2020, Chandra Imaging of the Western Hotspot in the Radio Galaxy Pictor A: Image Deconvolution and Variability Analysis. *ApJ*, 903(2), 109.

Wilson, A. S., Young, A. J., & Shopbell, P. L. 2000, Chandra Observations of Cygnus A: Magnetic Field Strengths in the Hot Spots of a Radio Galaxy. *ApJ Letters*, 544(1), L27–L30.

Worrall, D. M. 2009, The X-ray jets of active galaxies. *AApR*, 17, 1–46.

The Multimessenger Chakra of Blazar Jets
Proceedings IAU Symposium No. 375, 2023
I. Liodakis, M. F. Aller, H. Krawczynski, A. Lähteenmäki & T. J. Pearson, eds.
doi:10.1017/S1743921323000959

Multiwavelength monitoring of the nucleus in PBC J2333.9−2343: A giant radio galaxy with a Blazar-like core

L. Hernández-García[1,2], F. Panessa[3], G. Bruni[3], L. Bassani[4], P. Arévalo[2,5], V. M. Patiño-Alvarez[6,7], A. Tramacere[8], P. Lira[9,5], P. Sánchez-Sáez[10,1], F. E. Bauer[11,1,12], V. Chavushyan[6,13], R. Carraro[2], F. Förster[14,1,15], A. M. Muñoz Arancibia[1,15] and P. Ubertini[3]

[1]Millennium Institute of Astrophysics (MAS), Nuncio Monseñor Sótero Sanz 100, Providencia, Santiago, Chile
email: lorena.hernandez@uv.cl

[2]Instituto de Física y Astronomía, Facultad de Ciencias, Universidad de Valparaíso, Gran Bretaña No. 1111, Playa Ancha, Valparaíso, Chile

[3]INAF - Istituto di Astrofisica e Planetologia Spaziali, via Fosso del Cavaliere 100, I-00133 Roma, Italy

[4]INAF - Osservatorio di Astrofisica e Scienza dello Spazio, via P. Gobetti 101, I-40129 Bologna, Italy

[5]Millennium Nucleus on Transversal Research and Technology to Explore Supermassive Black Holes (TITANS)

[6]Instituto Nacional de Astrofísica, Optica y Electrónica, Luis Enrique Erro 1, Tonantzintla, Puebla 72840, México

[7]Max-Planck-Institut für Radioastronomie, Auf dem Hügel 69, D-53121 Bonn, Germany

[8]Department of Astronomy, University of Geneva, Ch. d' Écogia 16, Versoix, 1290, Switzerland

[9]Departamento de Astronomía, Universidad de Chile, Casilla 36D, Santiago, Chile

[10]European Southern Observatory, Karl-Schwarzschild-Str. 2, 85748, Garching, Germany

[11]Instituto de Astrofísica and Centro de Astroingeniería, Facultad de Física, Pontificia Universidad Católica de Chile, Casilla 306, Santiago 22, Chile

[12]Space Science Institute, 4750 Walnut Street, Suite 205, Boulder, CO 80301, USA

[13]Center for Astrophysics — Harvard & Smithsonian, 60 Garden Street, Cambridge, MA 02138, USA

[14]Data and Artificial Intelligence Initiative (IDIA), Faculty of Physical and Mathematical Sciences, University of Chile, Chile

[15]Center for Mathematical Modeling (CMM), Universidad de Chile, Beauchef 851, Santiago 8320000, Chile

Abstract. We present an observational multiwavelength campaign during 2018–19 for PBC J2333.9−2343, a giant radio galaxy with a bright central core associated to a blazar nucleus, whose structure could be due to a significant jet reorientation. We report flux increases by a factor of two or more on timescales shorter than a month, resembling flaring events. The cross correlation between the NIR and optical bands shows quasi-simultaneous variations arising from the jet. The optical variability properties of PBC J2333.9−2343 are more comparable to a sample of blazars than to non-blazar AGN. The SED of the nucleus shows two peaks, with a derived jet angle of 3 degrees, also typical of a blazar. Therefore, we confirm the presence of a blazar-like core in the center of this galaxy.

Keywords. galaxies: active, galaxies: nuclei, galaxies: variability

1. Introduction

Under the Unified Model of Active Galactic Nuclei (AGN), the different types of AGN are attributed to orientation effects, with radio galaxies showing a pair of jets in the plane of the sky, and blazars are sources where one of the jets is pointing to the observer (Urry & Padovani 1995).

However, one of the things that was not taken into account in this model was the fact that galaxies evolve with time, and can go through different phases of nuclear and/or jet activity. Reactivation and jet reorientation of AGN can be observed in some Giant Radio Galaxies (GRG, Ishwara-Chandra & Saikia 1999), which have projected linear sizes larger than 0.7 Mpc from radio-lobe to radio-lobe, and can have spectral ages of 10^7-10^8 yr (Alexander & Leahy 1987). X-shaped radio galaxies (XRG) are another possible consequence of reactivation or jet reorientation reported in the literature (Gillone, Capetti, & Rossi 2016). Sources of this kind are therefore perfect laboratories to study intermittent activity and AGN evolution because these processes occur on accessible timescales (i.e., the changes can be observed in a few years).

PBC J2333.9−2343 is a GRG whose Very Large Array (VLA) radio map shows two jets with a linear size of 1.2 Mpc (Bassani et al. 2016). Observations at milliarcsecond scale with the Very Long Baseline Array (VLBA) show an optically thick compact core with an optically thin jet. Hernández-García et al. (2017) studied these and *XMM–Newton* data and proposed that a Blazar-like nucleus beamed at an angle < 6 degrees to the line of sight resides at the center of this GRG. Moreover, Bruni et al. (2020) presented a deep image with the Giant Metrewave Radio Telescope (GMRT) at 150 MHz showing a lack of emission between the lobes and the core region. A study of its variability on timescales of years in the optical, UV and X-rays was presented in Hernández-García et al. (2018). To further constrain the properties of this peculiar galaxy, we performed a multiwavelength monitoring of its nucleus in 2018–19.

2. Data

We performed two monitoring campaigns, the first one between July 2018 and January 2019, and the second one between April-August 2019. The following data were used for the variability analysis as well as for building the spectral energy distribution (SED):

- Radio observations with the Effelsberg-100m telescope with a cadence of once per month at frequencies 4.8, 8.5, 10.5, and 20.4 GHz. For the SED we also used data from the Very Long Baseline Array (VLBA) at 15 and 24 GHz, Very Large Array Sky Survey (VLASS) at 3 GHz, and the Rapid ASKAP Continuum Survey at 0.88 GHz.
- Simultaneous near infrared and optical photometry with SMARTS-1.3m. Two campaigns were organized, in 2018 with 3–4 days cadence, and in 2019 with daily cadence.
- Optical photometry with the Zwicky Transient Facility (ZTF) and the Asteroid Terrestrial-impact Last Alert System (ATLAS) with 3–4 days cadences.
- Simultaneous X-ray (0.5-10 keV) and UV (UVM2 filter) data with the Neil Gehrels *Swift* Observatory with weekly cadence.
- Gamma-ray data from the Fermi Large Area Telescope (LAT). We were not able to obtain a light curve because the source was not detected with signal-to-noise ratio > 3 in bins of length smaller than one year. A detection at a 6σ level results from integrating

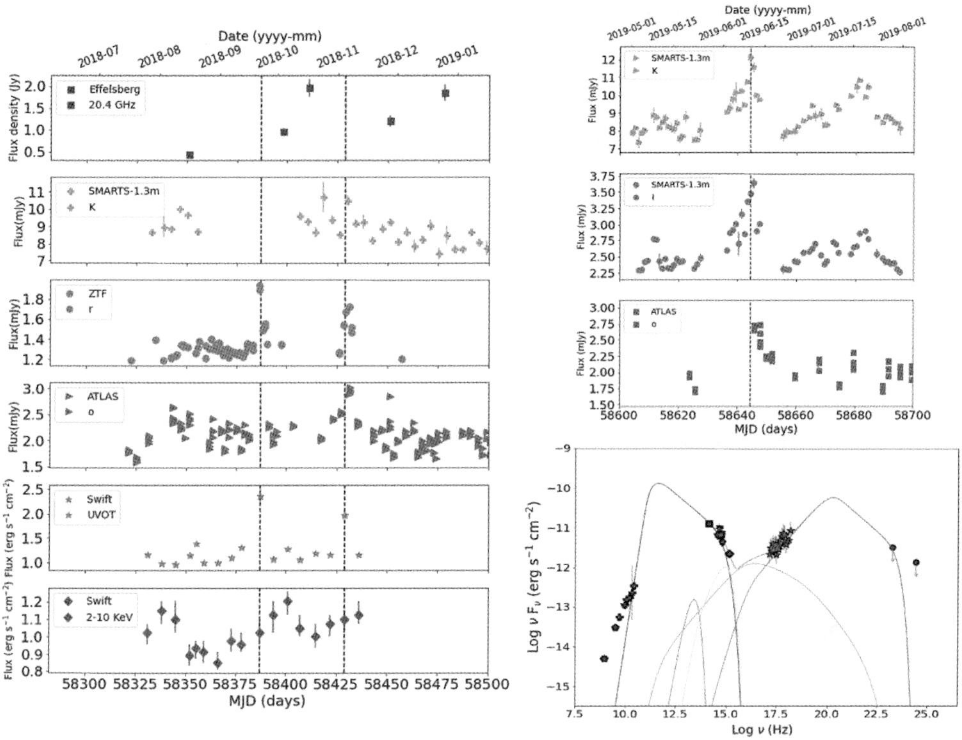

Figure 1. Left: Monitoring in 2018 from radio to X-rays (from top to bottom: Effelsberg, SMARTS-1.3m, ZTF, ATLAS, *Swift*/UVOT, and *Swift*/XRT); Upper-right: Monitoring in 2019 with SMARTS-1.3m and ATLAS; Lower-right: SED fitting using a leptonic model.

one year of data in the 0.1-300 GeV. For the SED we use upper limits in three equally spaced energy bins.

3. Main results

In Figure 1 (left) we plot the multiwavelength light curves for the monitoring in 2018, spanning between MJD=58320 (2018 July 21) and 58500 (2019 January 17). For simplicity and clarity, we plot only the most representative light curve per instrument. In the right panel we plot the monitoring performed between MJD=58600-58700 (April 27th–August 5th, 2019) with SMARTS-1.3m and ATLAS.

The results obtained from the analysis of these light curves and the multiwavelength data are presented in detail in Hernández-García et al. (2023), and here we report the main results:

• Variations are detected at all observed wavelengths at significance larger than 6σ, except at X-rays where variations are detected at 2σ within the four months.

• The observed flux variations show a flaring behaviour. Three events occurred during this monitoring and were detected at different frequencies.

• The cross-correlation between simultaneous optical/NIR shows a delay of 1.02 ± 1.45 days, i.e., suggesting that these variations occur quasi-simultaneously. The lack of delay is compatible with variations from the jet (see Hernández-García et al. 2023).

• When comparing the optical variability features with large samples of non-blazar AGN and blazars using samples from ALeRCE (Förster et al. 2021, Sánchez-Sáez et al. 2021), PBC J2333.9−2343 shows characteristics more similar to the blazar population.

• We constructed the SED using all the data described in Section 2, and it was then fitted with a single-zone leptonic model using the Jets SED modeler and fitting Tool (`JetSeT`, Tramacere 2020). The SED is presented in the right bottom panel of Fig.1, where two distinct peaks can be differentiated. The low energy peak is well fitted by synchrotron emission, while the high energy peak is dominated by External Compton (EC) from the dusty torus with some contribution from Synchrotron Self Compton (SSC). The jet angle is 3 degrees.

• We confirm the finding reported in Bruni et al. (2020) that there is no connection between the nucleus and the lobes using a deep RACS image.

• We confirm a very exceptional case of jet reorientation in PBC J2333.9−234, already proposed in Hernández-García et al. (2017), that shows a transformation from a GRG (with two lobes in the plane of the sky) to a blazar (with a jet pointing towards us).

4. Acknowledgements

We acknowledge funding from ANID programs: ICN12_009 (LHG, AMA, PSS, FEB, FF) and NCN19_058 (PA, PL); CATA-Basal - ACE210002 (FEB), FB210003 (FEB, FF) and BASAL project FB210005 (AMA); and FONDECYT Regular 1190818 (FEB) and 1200495 (FEB). From INTEGRAL ASI-INAF agreement 2019-35-HH.0 and ASI/INAF n. 2017-14-H.0 (GB, FP), from CONACyT research grants 280789 (VMPA, VC) and 320987 (VMPA), the MPIfR-Mexico (VMPA) and the MPA-UV (PA) Max Planck Partner Groups, the Fulbright - García Robles scholarship (VC), and the European Union's Horizon 2020 research and innovation program under grant agreement No. 730562 (RadioNet).

References

Alexander, P. & Leahy, J. P. 1987, *MNRAS*, 225, 1

Bassani L., Venturi T., Molina M., Malizia A., Dallacasa D., Panessa F., Bazzano A., & Ubertini P. 2016, *MNRAS*, 461, 3165

Bruni, G., Panessa, F., Bassani, L., Chiaraluce, E., Kraus, A., Dallacasa, D., Bazzano, A., Hernández-García, L., Malizia, A., Ubertini, P., Ursini, F., & Venturi, T. 2020, *MNRAS*, 494, 902

Förster, F., Cabrera-Vives, G., Castillo-Navarrete, E., Estévez, P. A., Sánchez-Sáez, P., Arredondo, J., Bauer, F. E., Carrasco-Davis, R., Catelan, M., Elorrieta, F., Eyheramendy, S., Huijse, P., Pignata, G., Reyes, E., Reyes, I., Rodríguez-Mancini, D., Ruz-Mieres, D., Valenzuela, C., Álvarez-Maldonado, I., Astorga, N., Borissova, J., Clocchiatti, A., De Cicco, D., Donoso-Oliva, C., Hernández-García, L., Graham, M. J., Jordán, A., Kurtev, R., Mahabal, A., Maureira, J. C., Muñoz-Arancibia, A., Molina-Ferreiro, R., Moya, A., Palma, W., Pérez-Carrasco, M., Protopapas, P., Romero, M., Sabatini-Gacitua, L., Sánchez, A., San Martín, J., Sepúlveda-Cobo, C., Vera, E., & Vergara, J. R. 2021, *AJ*, 161, 242

Gillone, M., Capetti, A., & Rossi, P. 2016, *A&A*, 587, A25

Hernández-García, L., Panessa, F., Giroletti, M., Ghisellini, G., Bassani, L., Masetti, N., Povic, M., Bazzano, A., Ubertini, P., Malizia, A., & Chavushyan, V. 2017, *A&A*, 603, A131

Hernández-García, L., Vietri, G., Panessa, F., Piconcelli, E., Chavushyan, V., Jiménez-Andrade, E. F., Bassani, L., Bazzano, A., Cazzoli, S., Malizia, A., Masetti, N., Monaco, L., Povic, M., Saviane, I., & Ubertini, P. 2018, *MNRAS*, 478, 4634

Hernández-García, L.; Panessa, F.; Bruni, G.; Bassani, L.; Arévalo, P.; Patiño-Alvarez, V. M.; Tramacere, A.; Lira, P.; Sánchez-Sáez, P.; Bauer, F. E.; Chavushyan, V.; Carraro, R.; Förster, F.; Muñoz Arancibia, A. M.; Ubertini, P. 2023, *MNRAS*, eprint arXiv:2303.08842

Ishwara-Chandra C. H. & Saikia D. J. 1999, *MNRAS*, 309, 100

Sánchez-Sáez, P., Reyes, I., Valenzuela, C., Förster, F., Eyheramendy, S., Elorrieta, F., Bauer, F. E., Cabrera-Vives, G., Estévez, P. A., Catelan, M., Pignata, G., Huijse, P., De Cicco, D.,

Arévalo, P., Carrasco-Davis, R., Abril, J., Kurtev, R., Borissova, J., Arredondo, J., Castillo-Navarrete, E., Rodriguez, D., Ruz-Mieres, D., Moya, A., Sabatini-Gacitúa, L., Sepúlveda-Cobo, C., & Camacho-Iñiguez, E. 2021, *AJ*, 161, 141
Tramacere, A. 2020, JetSeT: Numerical modeling and SED fitting tool for relativistic jets, Astrophysics Source Code Library, record ascl:2009.001 (ascl:2009.001)
Urry, C.M. & Padovani, P. 1995, *PASP*, 107, 803

The Multimessenger Chakra of Blazar Jets
Proceedings IAU Symposium No. 375, 2023
I. Liodakis, M. F. Aller, H. Krawczynski, A. Lähteenmäki & T. J. Pearson, eds.
doi:10.1017/S1743921323000698

A *WISE* perspective of the blazar hunt in the γ-ray sky

Francesco Massaro[1] and Raffaele D'Abrusco[2]

[1]Dipartimento di Fisica, Università degli Studi di Torino, via Pietro Giuria 1, I-10125 Torino, Italy

[2]Center for Astrophysics — Harvard & Smithsonian, 60 Garden Street, Cambridge, MA 02138, USA

Abstract. Thanks to the data of the *WISE* all-sky survey we discovered that the non-thermal infrared emission of blazars, the largest population of associated γ-ray sources, has peculiar spectral properties. Here we provide a summary of all results achieved on the *infrared–γ-ray connection*. We also show results on the latest statistical analysis of the tight correlation between the mid-infrared colors and the γ-ray spectral index for *Fermi* blazars, a connection that links both emitted powers and spectral shapes of particles accelerated in blazar jets over ten decades in frequency of the electromagnetic spectrum. Finally, we outline all developments performed in the last decade achieved using the infrared–γ-ray connection to discover hundreds of new blazars within the sample of unidentified γ-ray sources thanks to optical spectroscopic observations.

Keywords. active galactic nuclei, blazars, relativistic jets, surveys, gamma-rays

---

1. Introduction

Blazars are a class of radio-loud active galaxies showing an extreme behavior. Their emission spans the whole electromagnetic spectrum from radio to TeV energies exhibiting variability at all wavelengths with timescales ranging from weeks to minutes coupled with evidence of superluminal motions, high and variable polarization and flat radio spectra (see e.g. Urry & Padovani 1995; Massaro et al. 2009, for a review), even observed below $\sim$1 GHz (i.e., Massaro et al. 2013a,b).

Discoveries on the blazar population in the last decade where mainly achieved thanks by the survey carried out with the *Fermi* satellite (Atwood et al. 2009), since blazars constitute the dominant class of γ-ray sources. Blazars account for more than 1/3 of the *Fermi* detected sources (Acero et al. 2015). Thus we could also expect that a significant fraction of the unidentified/unassociated γ-ray sources (UGSs; Massaro et al. 2012, 2013c), in particular at high Galactic latitudes, also belong to this class.

Moreover, together with star forming galaxies (e.g. Ackermann et al. 2012) and radio galaxies (e.g. Di Mauro et al. 2014; Massaro & Ajello 2011), blazars produce a significant contribution to the extragalactic γ-ray background (Ajello et al. 2012, 2014; Massaro et al. 2016b, and references therein).

Blazars historically come in two flavors, mainly distinguished on the basis of their optical properties: BL Lac objects and flat spectrum radio quasars. According to the nomenclature introduced by the *Roma-BZCAT* (Massaro et al. 2015b) The former subclass is generally labelled as BZBs while the latter as BZQs. In particular, BZBs show featureless optical spectra and/or with weak absorption lines of equivalent width lower than 5$\AA$, while the BZQs have a typical quasar-like optical spectra (Stickel et al. 1991; Falomo et al. 2014).

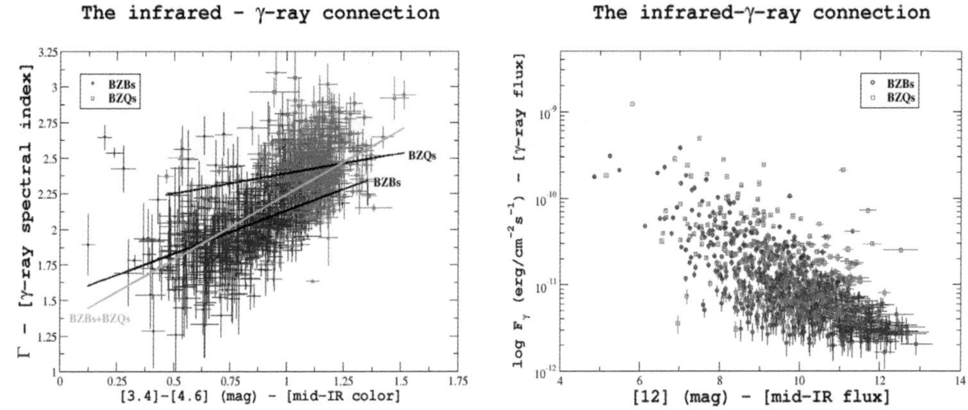

Figure 1. (Left panel) The γ-ray spectral index Γ *vs* the mid-IR color [3.4]-[4.6] μm for the blazars belonging to the *WISE locus*. BZBs are shown in blue open circles while BZQs in red open squares. Linear correlations between the γ-ray spectral and the mid-IR shaped are highlighted by regression lines in black for BZBs and BZQs while in orange for the whole population, respectively. This is the observational evidence of the infrared–γ-ray connection. **(Right panel)** The *WISE* magnitude at [12] μm *vs* the γ-ray flux density for all blazars included in the to the *WISE locus*. Here we report the trend between fluxes using the magnitude at [12] μm, being less contaminated by the emission arising from the host galaxy than that measured at 3.4 μm, and the γ-ray flux. Similar trends/correlations exists also using all other *WISE* magnitudes (see e.g., Massaro & D'Abrusco 2016, for all details about the analysis).

Starting from 2010, the NASA Wide-field Infrared Survey Explorer (*WISE*; Wright et al. 2010) mapped the sky at mid-infrared (mid-IR) frequencies, in 4 bands at nominal wavelenghts: 3.4, 4.6, 12, and 22 μm, making possible the statistical investigation of the mid-IR properties of the blazar population.

In 2011, we discovered that *Fermi* blazars inhabit a region of the mid-IR color-color diagram (i.e., [3.4]-[4.6]-[12] μm), built with the *WISE* magnitudes, well separated from the location of other extragalactic sources (Massaro et al. 2011; D'Abrusco et al. 2012). This 2-dimensional region of the mid-IR color-color diagram was originally labelled as the *WISE Gamma-ray Strip*. Then we found that the *WISE Gamma-ray Strip* is the projection of a 3-dimensional volume, in the mid-IR color space, today known as the *WISE locus* of γ-ray blazars (D'Abrusco et al. 2013, 2014). Here we show the results achieved on the link found between the *WISE* mid-IR colors and the γ-ray photon index as well as between fluxes measured in these energy ranges (Massaro & D'Abrusco 2016).

We use cgs units unless stated otherwise. Gamma-ray photon index, Γ is defined by the usual convention on the flux density, $N(E) \propto E^{-\Gamma}$, being $N(E)$ the number of γ-ray photons detected per unit of time, area and energy. *WISE* magnitudes are in the Vega system and are not corrected for the Galactic extinction since such correction affects significantly only the magnitude at 3.4 μ for sources lying at low Galactic latitudes (see e.g. D'Abrusco et al. 2014).

2. On the infrared–γ-ray connection

To assess the significance of the *infrared–γ-ray connection*, we searched a correlation between the γ-ray photon index Γ and their *WISE* mid-IR colors, a surrogate of the mid-IR spectral index, for all *Fermi* blazars. We discovered a strong correlation, the first evidence of the existence of the *infrared–γ-ray connection*, between these two parameters, as reported in Fig. 1 (see Massaro & D'Abrusco 2016, for whole details about the statistical analysis).

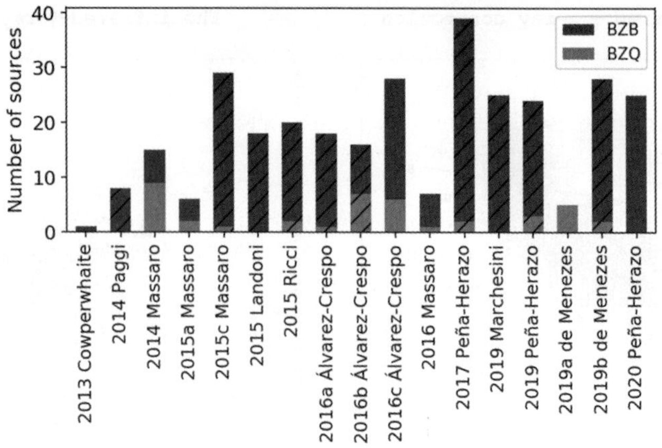

Figure 2. Number of blazars discovered during the optical spectroscopic campaign carried out pointing γ-ray blazar candidates that lie within the positional γ-ray uncertainty region and have mid-IR colors similar to that of blazars (hatched bars). We also show blazars discovered thanks to archival searches on spectroscopic surveys (non-hatched bars). BZBs are shown in blue while BZQs in red as previously labelled.

We first searched for a correlation between spectral shapes in the Γ *vs* [3.4]-[4.6] μm plane. The correlation we found appears to be more statistically significant for BZBs rather than for BZQs with a linear correlation coefficient $\rho = 0.47$ for the former ones and a negligible probability to be spurious, given the large datasets used. We obtained $\rho = 0.15$ for the BZQs computed using a sample of 419 γ-ray sources that also in this case led to a negligible chance probability. When combining datasets of both classes, for a total of 1022 *Fermi* blazars, the linear correlation coefficient, unsurprisingly, increases to $\rho = 0.65$. Thus the slope of the best fit regression lines, as expected, is mainly driven by the stronger correlation found on the sources belonging to the BZB class.

A similar situation occurs in the Γ *vs* [4.6]-[12] μm plane, where the correlation coefficients for the 595 BZBs and the 419 BZQs separately are $\rho = 0.48$ and $\rho = 0.13$, respectively. Then the correlation coefficient computed for the whole *Fermi* blazar sample, in this case listing 1014 γ-ray sources is $\rho = 0.60$, lower than the previous one due to the larger scatter induced by the uncertainties of mid-IR magnitudes at 12μm larger than those at 3.4 μm and 4.6 μm.

For the sake of comparison, it is worth stressing that the *radio–γ-ray connection* is based on the correlation between the radio and γ-ray flux densities. A similar trend also exists for the mid-IR emission as shown in Fig. 1, which displays the 12 μm magnitude vs γ-ray flux density for both *Fermi* BZBs and BZQs.

3. Summary and conclusions

Several years after the discovery that the non-thermal emission of γ-ray blazars can be traced using mid-IR colors obtained from the photometry of the *WISE* all-sky survey, we report the latest results on the existence of a *infrared–γ-ray connection* for the *Fermi* blazars that appears to be at least as strong as the well-known radio–γ-ray one.

We found statistically significant correlations between the γ-ray photon index and the mid-IR colors for both the whole sample of *Fermi* blazars and the BZBs and BZQs spectral classes separately, the basis of the *infrared–γ-ray connection.* This correlation appears "stronger" than the well known radio–γ-ray connection because it involves the spectral shapes of the *Fermi* blazars over $\sim$10 orders of magnitude and not only their flux densities.

Taking advantage of the overwhelming fraction of *Fermi* blazars detected in the *WISE* all-sky survey (i.e., ∼99%) in the first two mid-IR filters, we suggested a comprehensive investigation of their IR properties at the light of the *infrared–γ-ray connection* since it has been a powerful tool to reveal the real fraction of *Fermi* blazars hidden within the sample of UGSs. To emphasize its use we finally, report in Fig. 2, a summary of the spectroscopic campaign carried out in the last decade to observed γ-ray blazar candidates that lie within the positional γ-ray uncertainty region and have mid-IR colors similar to that of blazars. This was also augmented by an extensive archival search in the databases of the major optical surveys that significantly helped to discover new γ-ray blazars as also shown in Fig. 2.

References

Acero, F., Ackermann, M., Ajello, M. et al. 2015 ApJS, 218, 23
Ackermann, M., Ajello, M., Allafort, A. et al. 2012a ApJ, 755, 164
Ajello, M., Shaw, M. S., Romani, R. W. et al. 2012 ApJ, 751, 108
Ajello, M. et al. 2014 ApJ, 780, 73
Álvarez Crespo, N., Masetti, N., Landoni, M. et al. 2016a, AJ, 151, 32
Álvarez Crespo, N., Massaro, F., Milisavljevic, D. et al. 2016b, AJ, 151, 95
Álvarez-Crespo, N., Massaro, F., D'Abrusco, R. et al. 2016c, Ap&SS 361, 316
Atwood, W. B., Abdo, A. A., Ackermann, M. et al. 2009 ApJ, 697, 1071
Cowperthwaite, Philip S. et al. 2013 AJ, 146, 110
D'Abrusco, R., Massaro, F., Ajello, M. et al. 2012 ApJ, 748, 68
D'Abrusco, R., Massaro, F., Paggi, A. et al. 2013 ApJS, 206, 12
D'Abrusco, R., Massaro, F., Paggi, A. et al. 2014 ApJS, 215, 14
de Menezes, R., Peña-Herazo, H. A., Marchesini, E. J., et al. 2019 A&A 630, A55
de Menezes, R., Amaya-Almazán, R. A., Marchesini, E. J., et al. 2020 ApSS, 365, 12
Di Mauro, M., Calore, F., Donato, F., Ajello, M., Latronico, L. 2014 ApJ, 780, 161
Falomo, R., Pian, E., Treves, A. 2014 A&ARv, 22, 73
Landoni, M., Massaro, F., Paggi, A., et al. 2015a AJ, 149, 163
Marchesini, E. J., Peña-Herazo, H. A., Álvarez Crespo, N. et al., 2019, Ap&SS, 364, 5
Massaro, E., Giommi, P., Leto, C. et al. 2009 A&A, 495, 691
Massaro, F. & Ajello, M. 2011b ApJ, 729L, 12
Massaro, F., D'Abrusco, R., Ajello, M. et al. 2011a ApJ, 740L, 48
Massaro, F., D'Abrusco, R., Tosti, G. et al. 2012a ApJ, 752, 61
Massaro, F., D'Abrusco, R., Giroletti, M. et al. 2013a ApJS, 207, 4
Massaro, F., Giroletti, M., Paggi, A. et al. 2013b ApJS, 208, 15
Massaro, F., D'Abrusco, R., Paggi, A. et al. 2013c ApJS, 206, 13
Massaro, F., Masetti, N., D'Abrusco, R. et al. 2014, AJ, 148, 66
Massaro, E., Maselli, A., Leto, C. et al. 2015b Ap&SS, 357, 75
Massaro, F., D'Abrusco, R., Landoni, M. et al. 2015a, ApJS, 217, 2
Massaro, F., Landoni, M., D'Abrusco, R. et al. 2015c, A&A, 575, 124
Massaro, F., Álvarez Crespo, N., D'Abrusco, R. et al. 2016, Ap&SS 361, 337
Massaro, F. & D'Abrusco, R. 2016, ApJ 827, 67
Massaro, F., Thompson, D. J., Ferrara, E. C. 2015 A&AR, 24, 2
Paggi, A., Milisavljevic, D., Masetti, N. et al. 2014, AJ, 147, 112
Peña-Herazo, H. A., Marchesini, E. J., Álvarez Crespo, N. et al. 2017 Ap&SS, 362, 228
Peña-Herazo, H. A., Massaro, F., Chavushyan, V., et al. 2019 ApSS, 364, 85
Peña-Herazo, H. A., Amaya-Almazán, R. A., Massaro, F. et al. 2020 A&A, 643, 103
Ricci F., Massaro F., Landoni M. et al. 2015 AJ, 149, 160
Stickel, M., Padovani, P., Urry, C. M. et al. 1991 ApJ, 374, 43
Urry, C. M., & Padovani, P. 1995, PASP, 107, 803
Wright, E. L., et al. 2010 AJ, 140, 1868

The Multimessenger Chakra of Blazar Jets
Proceedings IAU Symposium No. 375, 2023
I. Liodakis, M. F. Aller, H. Krawczynski, A. Lähteenmäki & T. J. Pearson, eds.
doi:10.1017/S1743921323001242

Gamma-ray flux distribution analysis on 145 gamma-ray bright blazars

Kenji Yoshida[1], Foteini Oikonomou[2], Maria Petropoulou[3] and Kohta Murase[4]

[1]Shibaura Institute of Technology, Japan

[2]Norwegian University of Science and Technology, Norway

[3]National and Kapodistrian University of Athens, Greece

[4]The Pennsylvania State University, USA

Abstract. We present the results of the gamma-ray flux distribution analysis on 145 gamma-ray bright blazars observed by Fermi-LAT. For the gamma-ray flux distribution, we applied a log-normal distribution to discuss the nature of the high-energy emission processes of blazars and a power-law distribution convolved with a Poisson distribution to investigate the implications of gamma-ray bright blazars for neutrino emission. Both distributions can represent the observed flux distributions as well. The leptonic models, which give the physical relationship between neutrinos and gamma rays, indicate that the flaring contribution to the neutrino emission can be dominant for the power-law index less than ~ 2.5. From the power-law distribution analysis, we found that the power-law index < 2.5 accounts for the 82 % blazars. This result suggests that the flaring contribution of blazars is dominant for high-energy neutrino emission.

Keywords. galaxies: active, galaxies: jets, gamma rays: observations, methods: data analysis

1. Introduction

The flux distributions of blazars have been used to study the temporal variation to provide valuable clues to the origin and nature of their variability. One distribution of light curves which is commonly found in blazars is the log-normal distribution (e.g., Shah *et al.* 2018; Bhatta *et al.* 2020). The log-normal distribution can interpret as multiplicative processes with the analog of the normal distribution as additive processes (Uttley *et al.* 2005, and references therein). That is, a process produced by the multiplication of many independent ones may have a log-normal distribution. On the other hand, Scargle (2020) discusses that a distribution resembling a log-normal one can be reproduced by additive processes with a power-law distribution.

As another approach, Murase *et al.* (2018) applied a power-law distribution convolved with a Poisson distribution for six blazar light curves using the public Fermi All-sky Variability Analysis (FAVA) data by *Fermi* LAT (Abdollahi *et al.* 2017). As described in Murase *et al.* (2018), the gamma-ray flux distribution analysis provides us how dominant the flaring contribution is to the neutrino emission, according to the various leptonic models, which give the physical relationship between neutrinos and gamma rays.

In this paper, we present a statistical study of the gamma-ray flux distribution analysis to investigate the implications of blazars for high-energy neutrino emission.

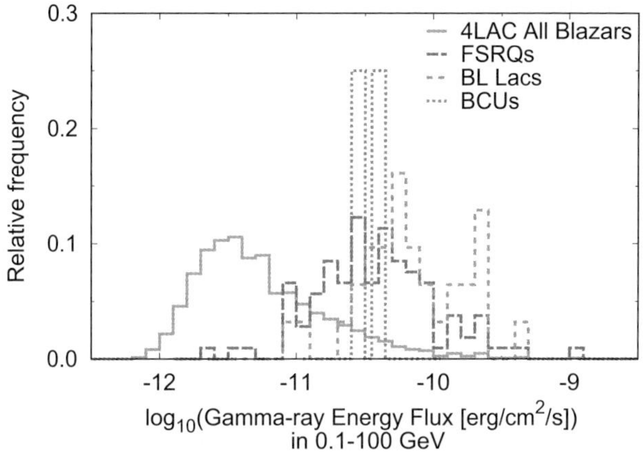

Figure 1. Relative frequencies of the $0.1 - 100$ GeV gamma-ray energy fluxes of 4LAC all 1701 blazars (dark gray line), 106 FSRQs (blue line), 31 BL Lacs (orange line), and eight BCUs (green line).

2. Data analysis

Our samples are 145 gamma-ray bright blazars observed by Fermi-LAT. The blazar subclasses are 106 FSRQs, 31 BL Lacs, and eight blazar candidates of uncertain type (BCUs). Figure 1 shows relative frequencies of the gamma-ray energy fluxes in $0.1 - 100$ GeV of our samples with 4LAC all 1701 blazars (Ajello *et al.* 2020). As shown in Fig. 1, our sample blazars are bright gamma-ray blazars in 4LAC blazars.

Using the $0.1 - 316$ GeV gamma-ray light curves with a one-week time bin through 2008-2019 in Yoshida *et al.* (2023), we derived the distribution of the number of time bins with gamma-ray flux: dN/dF_γ. We applied two model distributions for the gamma-ray flux distributions: log-normal and power-law ones. A log-normal distribution is given by

$$\frac{dN}{dF_\gamma} = \frac{N}{\sqrt{2\pi}F_\gamma \sigma} \exp\Big(-\frac{(\ln F_\gamma - \mu)^2}{2\sigma^2}\Big), \tag{1}$$

where N is the total number of time bins, μ and σ are the mean and standard deviation in units of the log of gamma-ray flux F_γ. A power-law distribution convolved with a Poisson distribution is given by

$$\frac{dN}{dk} = \sum_{n=n_0}^{\infty} An^{-\alpha} P(n + n_B, k + n_B), \tag{2}$$

where A is a normalization factor, k is the number of detected photons with a time bin, n and n_B are the numbers of source and background photons, $P(n + n_B, k + n_B)$ is a Poisson distribution, and n_0 is a minimum number of photons per bin that corresponds to a "quiescent" flux. Figure 2 gives two examples of the gamma-ray flux distributions fitted with a log-normal distribution and a power-law distribution convolved with a Poisson distribution for PKS 0250-225 and Mkn 501.

The Kolmogorov-Smirnov goodness-of-fit test gives the p-values under the null hypothesis that the observed flux distribution follows a model flux distribution. The blazars with a p-value > 0.05 account for 83% for the log-normal distribution and 86% for the power-law distribution. Hence the power-law distribution is applicable in equal or more blazars than the log-normal distribution. Figure 3 presents scatter plots between the log-normal mean μ and the power-law quiescent flux F_0, and the log-normal standard deviation σ

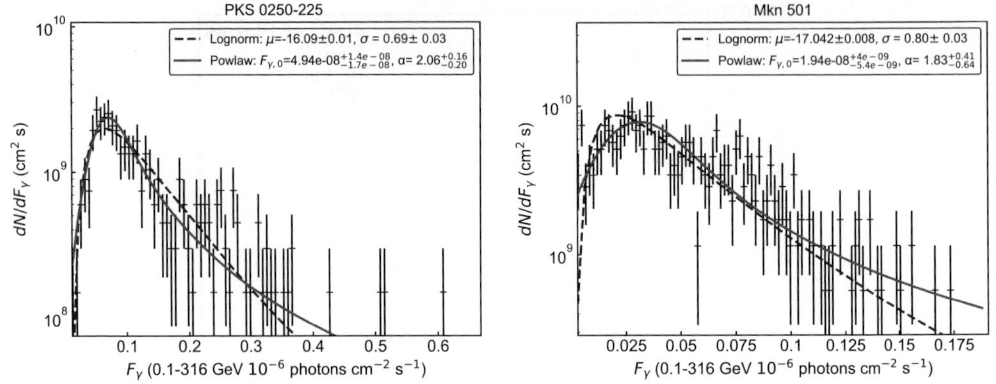

Figure 2. The gamma-ray flux distributions of PKS 0250-225 (left) and Mkn 501 (right) in $0.1-316$ GeV fitted with a log-normal distribution (blue line) and a power-law distribution convolved with a Poisson distribution (red line).

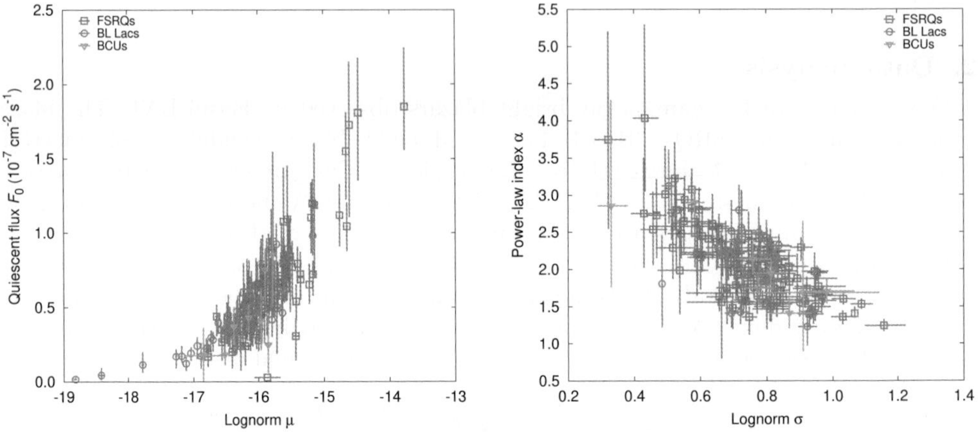

Figure 3. Scatter plots between the log-normal mean μ and power-law quiescent flux F_0 (left), and the log-normal standard deviation σ and power-law index α (right). The blue squares, orange circles, and green triangles show FSRQs, BL Lacs, and BCUs, respectively.

and the power-law index α. As we might expect, there are clear correlations between μ and F_0, and between σ and α as shown in Fig. 3.

As for μ and F_0, the Brunner-Munzel test (Brunner & Munzel 2000) suggests significant differences between FSRQs and BL Lacs with p-values of 5.3×10^{-5} % for μ and 1.9×10^{-2} % for F_0 for the null hypothesis that FSRQs and BL Lacs are equal. On the other hand, the Brunner-Munzel test suggests no significant differences between FSRQs and BL Lacs with p-values of 32 % for the log-normal standard deviation σ and 39 % for the power-law index α.

3. Discussion and Conclusions

The high-energy neutrinos from blazar jets are expected to be produced by photohadronic interactions of high-energy protons with target photons. Under different assumptions of the target photons, the leptonic models give a relation between the neutrino luminosity and gamma-ray luminosity of $L_\nu \propto L_\gamma^\gamma$ with $\gamma \simeq 1.0-2.0$ (Murase & Waxman 2016; Murase *et al.* 2018). Taking the power-law flux distribution

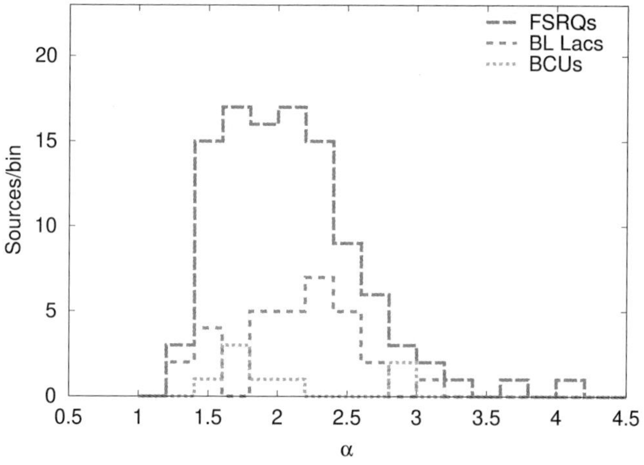

Figure 4. Distributions of power-law index α for FSRQs (blue line), BL Lacs (orange line), and BCUs (green line). The power-law index α less than 2.5 accounts for the 82% (119/145) blazars.

of $dN/dL_\gamma \propto {L_\gamma}^{-\alpha}$, we obtain

$$ {L_\nu}^2 \frac{dN}{dL_\nu} \propto {L_\nu}^{\frac{\gamma+1-\alpha}{\gamma}}. \tag{3} $$

This relation implies that gamma-ray flares can dominate the neutrino production of a blazar for $\alpha < \gamma + 1$. Using $dN/dL_\gamma \propto dN/dF_\gamma$ and $dN/dL_\nu \propto dN/dF_\nu$, we can estimate the power-law index α from the gamma-ray flux distributions with a power-law distribution convolved with a Poisson distribution. Figure 4 presents frequency distributions of power-law index α for FSRQs, BL Lacs, and BCUs. Corresponding to $\alpha < \gamma + 1$, the power-law indices α less than 2.0 with $\gamma = 1.0$, 2.5 with $\gamma = 1.5$, and 3.0 with $\gamma = 2.0$ account for the 46 % (67/145), 82 % (119/145), and 96 % (139/145) blazars, respectively. Hence, our results suggest that the flaring contribution of gamma-ray bright blazars is dominant for high-energy neutrino emission. The power-law index $\alpha = 2.1 \pm 0.3$ of TXS 0506+056 might be the case for the multi-messenger flare associated with IceCube-170922A (IceCube-Collaboration *et al.* 2018).

References

Abdollahi, S., Ackermann, M., Ajello, M., *et. al* 2017, *ApJ*, 846, 34
Ajello, M., Angioni R., Axelsson M., *et. al* 2020, *ApJ*, 892, 105
Bhatta, G. & Dhital, N. 2020, *ApJ*, 891, 120
Brunner E. & Munzel U. 2000, *Biometrical Journal*, 42, 17
IceCube-Collaboration, Fermi-LAT, MAGIC *et al.* 2018, *Sci.*, 361, 146
Murase, K. & Waxman, E. 2016, *Phys. Rev. D*, 94, 103006
Murase, K., Oikonomou, F., & Petropoulou, M. 2018, *ApJ*, 865, 124
Scargle, J.D. 2020, *ApJ*, 895, 90
Shah, Z., Mankuzhiyil, N., Sinha, A. *et al.* 2018, *Res. Astron. Astrophys.*, 18, 141
Uttley, P., McHardy, M., & Vaughan, S. 2005, *Mon. Not. R. Astron. Soc.*, 359, 345
Yoshida, K., Petropoulou, M., Murase, K., & Oikonomou, F. 2023, *ApJ*, submitted, arXiv:2210.10011

The Multimessenger Chakra of Blazar Jets
Proceedings IAU Symposium No. 375, 2023
I. Liodakis, M. F. Aller, H. Krawczynski, A. Lähteenmäki & T. J. Pearson, eds.
doi:10.1017/S1743921323000777

Milliarcsecond Core Size Dependence of the Radio Variability of Blazars

Po-Chih Hsu[1,2], Jun Yi Koay[1], Satoki Matsushita[1], Chorng-Yuan Hwang[2], Talvikki Hovatta[3,4], Sebastian Kiehlmann[5,6,7], Walter Max-Moerbeck[8], Tim Pearson[5], Anthony Readhead[5], Rodrigo Reeves[9] and Harish Vedantham[10,11]

[1]Institute of Astronomy and Astrophysics, Academia Sinica, Taipei 10617, Taiwan (R.O.C.)

[2]Graduate Institute of Astronomy, National Central University, Taoyuan 32001, Taiwan (R.O.C.)

[3]Finnish Centre for Astronomy with ESO (FINCA), University of Turku, FI-20014 Turku, Finland

[4]Aalto University Metsähovi Radio Observatory, Metsähovintie 114, FI-02540 Kylmälä, Finland

[5]Owens Valley Radio Observatory, California Institute of Technology, Pasadena, CA 91125, USA

[6]Institute of Astrophysics, Foundation for Research and Technology-Hellas, GR-71110 Heraklion, Greece

[7]Department of Physics, University of Crete, GR-70013 Heraklion, Greece

[8]Departamento de Astronomía, Universidad de Chile, Camino El Observatorio 1515, Las Condes, Santiago, Chile

[9]Departamento de Astronomía, Universidad de Concepción, Concepción, Chile

[10]Cahill Center for Astronomy and Astrophysics, California Institute of Technology, 1200 E. California Blvd., Pasadena, CA 91125, USA

[11]Netherlands Institute for Radio Astronomy (ASTRON), Oude Hogeveensedijk 4, NL-7991 PD Dwingeloo, the Netherlands

Abstract. Studying blazar radio variability on timescales ranging from months to years provides information on the sub-parsec-scale structures of the jets, and the physics of the central active galactic nuclei. In this study, we focus on the radio variability of 1158 blazars observed at 15 GHz through the Owens Valley Radio Observatory Blazar Monitoring Program, where these sources have been observed about twice a week for over a decade. We investigate the dependence of the variability amplitudes and timescales, derived using a simple model fit to the structure function, on the milliarcsecond radio core sizes measured by Very Long Baseline Interferometry. The most compact sources exhibit larger variability amplitudes and shorter variability timescales than the more extended sources; this could be explained by light travel-time effects.

Keywords. galaxies: active – galaxies: jets – quasars: general

1. Introduction

Radio variability of active galactic nuclei (AGNs) provides an opportunity to study the physics of the most compact regions of super massive black holes. Blazar variability can originate from the shocks produced along the jet (Hughes et al 1985; Hovatta et al 2008), jet precession (Kudryavtseva et al 2011), variable accretion rates due to disk instabilities (Lin and Shields 1986), or tidal disruption events (Alexander et al 2020), among others causes. Blazars, a subclass of AGN with relativistic jets aligned at a small angle to our line of sight, are suitable for such studies due to their compact and variable nature.

In this study, we used radio variability data from the 15 GHz Owens Valley Radio Observatory (OVRO) blazar monitoring program (Richards et al 2011), where more than 1500 blazars have been observed for over a decade, providing a large blazar sample with good time sampling (typical cadence of about 4 days). We used 1158 sources selected from the Candidate Gamma-Ray Blazar Survey (CGBaBS, Healey et al 2008). We also make use of multi-frequency blazar core sizes (1-86 GHz) derived by Koryukova et al (2022), which extends the work of an earlier paper by Pushkarev et al (2015). The core sizes were derived by fitting two circular Gaussian component models (Kovalev et al 2005) to the self-calibrated VLBA visibilities for each of their sources, with the most compact component being considered to be the core component.

We cross match the OVRO sample with the Koryukova et al (2022) sample with core size measurements. There are 116, 865, 464, 1053, 376, 156, 158, and 1 OVRO source(s) with available core size measurements at 1, 2, 5, 8, 15, 24, 43, and 86 GHz, respectively. In this study, we used angular core sizes from 2 to 43 GHz to investigate the dependency of variability amplitudes and timescales on the core sizes.

2. Analysis

We used the structure function (SF) to characterize the variability of the sources. The observed structure function $D_{\rm obs}(\tau)$ at a given time lag τ is given by:

$$D_{\rm obs}(\tau) = \frac{1}{N_\tau} \sum_{j,k} [S(t_{\rm j}) - S(t_{\rm k} - \tau)]^2 \tag{1}$$

where $S(t)$ is the measured flux density at time t, normalized by the mean flux density, and N_τ is the number of pairs of flux densities with a time lag τ. Next, we fit a simple model to the SF, assuming the variations are stochastic in nature, given by,

$$D_{\rm mod}(\tau) = D(1000{\rm d}) \frac{1 + \tau_{\rm char}/1000}{1 + \tau_{\rm char}/\tau} + {\rm D}_{\rm noise} \tag{2}$$

where $D(1000{\rm d})$ is the SF amplitude at 1000 days, $\tau_{\rm char}$ is the characteristic variability timescale, and $D_{\rm noise}$ is a constant additive noise floor due to systematic and instrumental errors that contribute to the overall variability.

For our analysis, we use the SF amplitude at 1000 days, derived from the above model fit, to characterize the long term variability amplitudes. We define the characteristic timescale of variability, as the time lag at which the SF amplitude reaches half the maximum of the saturation value.

3. Results and Discussion

3.1. *Dependence of Variability Amplitude on Core Sizes*

To test the dependence of variability amplitudes at 1000 days with angular core sizes, we separate the sources into strong ($\theta_{\rm strong}$) and weak variables ($\theta_{\rm weak}$), based on the median value of the variability amplitudes at 1000 days ($\tilde{D}(1000{\rm d})$). We then compare the distributions of the angular core sizes of the strong (blue histograms) and weak (orange histograms) variables, as shown in figure 1. We performed the two sample K-S test (results presented in table 1 test A), and found that weak variable sources (orange histograms) have significantly larger angular core sizes than strong variable (blue histograms) sources.

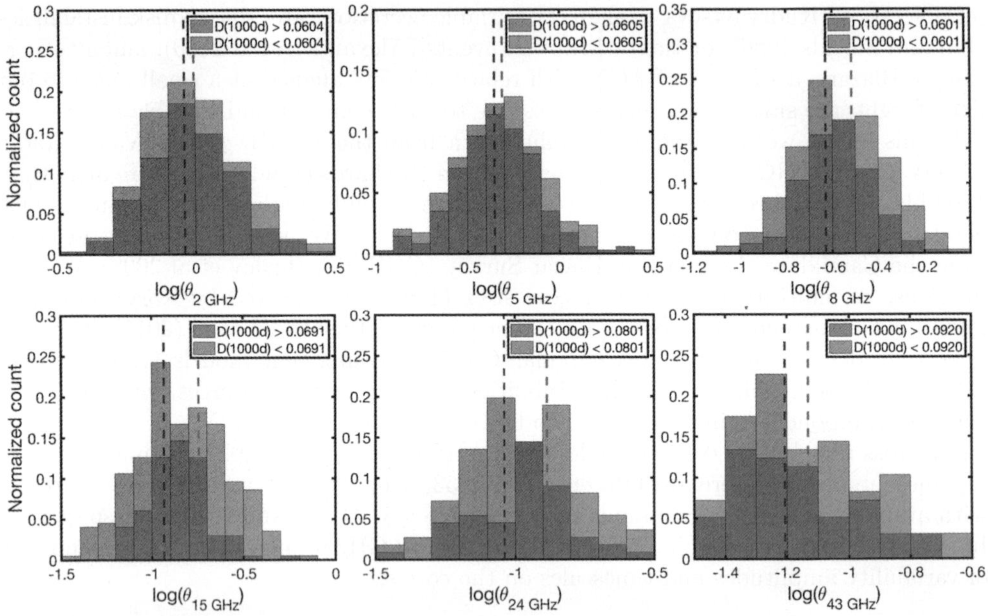

Figure 1. Normalized distributions of angular core sizes (θ) at different frequencies, separated into strong and weak variables by the median of the 15 GHz variability amplitudes D(1000d). The blue and orange dashed lines show the median value of the angular core sizes of the strong and weak variables, respectively. The brown areas correspond to overlapping regions between the blue and orange histograms.

Table 1. Results of two sample K-S test. Test A is the K-S test comparing the distribution of angular core sizes of weak and strong variables at 6 frequencies; test B is the K-S test comparing the distribution of angular core sizes of slow and fast variables at 6 frequencies. Sample sizes, median variability amplitudes at 1000 days $\tilde{D}$(1000d) or median of characteristic timescale $\tilde{\tau}_{\rm char}$, median core sizes of each distribution, as well as p value of each test are given in the table.

Test A (Separated by $\tilde{D}$(1000d))	**Sample number**	**$\tilde{D}$(1000d)**	**$\tilde{\theta}_{\rm weak}$ [mas]**	**$\tilde{\theta}_{\rm strong}$ [mas]**	**p value**
Distribution of $\theta_{\rm 2\ GHz,weak}$ and $\theta_{\rm 2\ GHz,strong}$	865	0.0604	0.99	0.91	1.83×10^{-2}
Distribution of $\theta_{\rm 5\ GHz,weak}$ and $\theta_{\rm 5\ GHz,strong}$	464	0.0605	0.49	0.45	1.30×10^{-1}
Distribution of $\theta_{\rm 8\ GHz,weak}$ and $\theta_{\rm 8\ GHz,strong}$	1053	0.0601	0.31	0.24	7.77×10^{-17}
Distribution of $\theta_{\rm 15\ GHz,weak}$ and $\theta_{\rm 15\ GHz,strong}$	376	0.0691	0.18	0.12	2.58×10^{-15}
Distribution of $\theta_{\rm 24\ GHz,weak}$ and $\theta_{\rm 24\ GHz,strong}$	156	0.0801	0.13	0.09	1.58×10^{-6}
Distribution of $\theta_{\rm 43\ GHz,weak}$ and $\theta_{\rm 43\ GHz,strong}$	158	0.0920	0.07	0.06	5.78×10^{-2}
Test B (Separated by $\tilde{\tau}_{\rm char}$)	**Sample number**	**$\tilde{\tau}_{\rm char}$ [days]**	**$\tilde{\theta}_{\rm slow}$ [mas]**	**$\tilde{\theta}_{\rm fast}$ [mas]**	**p value**
Distribution of $\theta_{\rm 2\ GHz,slow}$ and $\theta_{\rm 2\ GHz,fast}$	965	650 d	0.99	0.91	4.72×10^{-4}
Distribution of $\theta_{\rm 5\ GHz,slow}$ and $\theta_{\rm 5\ GHz,fast}$	464	637 d	0.50	0.44	1.78×10^{-1}
Distribution of $\theta_{\rm 8\ GHz,slow}$ and $\theta_{\rm 8\ GHz,fast}$	1053	630 d	0.25	0.28	1.07×10^{-4}
Distribution of $\theta_{\rm 15\ GHz,slow}$ and $\theta_{\rm 15\ GHz,fast}$	376	560 d	0.16	0.12	2.33×10^{-6}
Distribution of $\theta_{\rm 24\ GHz,slow}$ and $\theta_{\rm 24\ GHz,fast}$	156	544 d	0.12	0.10	4.52×10^{-2}
Distribution of $\theta_{\rm 43\ GHz,slow}$ and $\theta_{\rm 43\ GHz,fast}$	158	482 d	0.07	0.06	6.33×10^{-2}

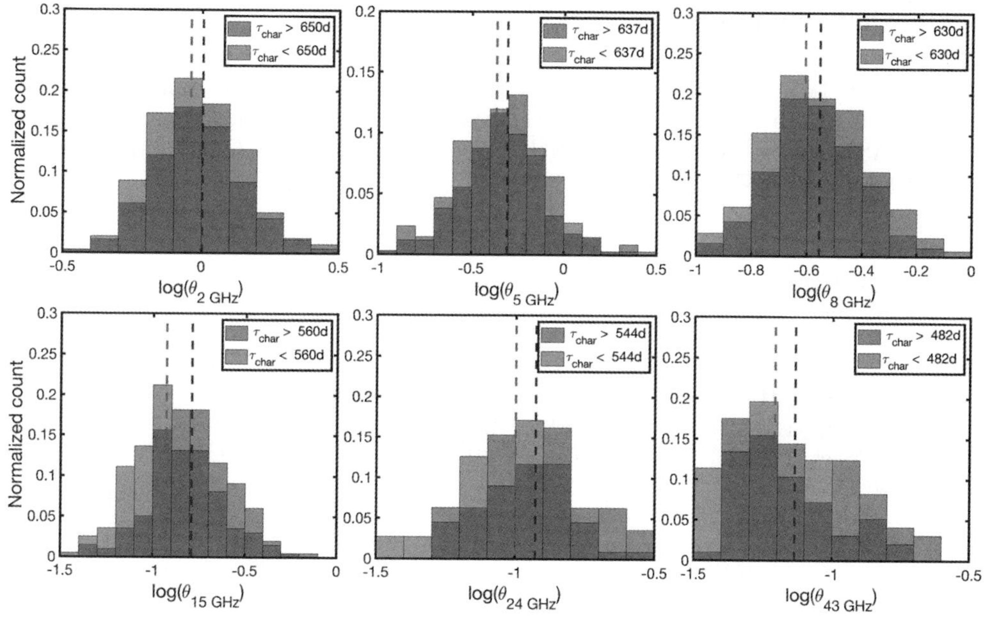

Figure 2. Normalized distributions of angular core size (θ) of slow and fast variables, separated by the median of characteristic timescales τ_{char}. The blue and orange dashed lines show the median of angular core sizes of the slow and fast variables at each frequency. The brown areas correspond to overlapping regions between the blue and orange histograms.

The p-value is larger at 15 GHz compared to that at 8 GHz, likely due to the smaller sample size.

3.2. *Dependence of Variability Timescale on Core Sizes*

For each frequency at which the core size measurements are available, we separate the source sample into fast (θ_{fast}) and slow variables (θ_{slow}), based on the median of the characteristic timescale ($\tilde{\tau}_{\text{char}}$). Figure 2 displays the distribution of the angular sizes of the fast (orange histograms) and slow (blue histograms) variables. We also perform two sample K-S tests comparing the distributions. We found that for core sizes measured at 2, 8, 15, and 24 GHz, the fast variable sources have significantly smaller angular core sizes than slow variable sources (see table 1 test B).

4. Discussion and Conclusion

We found significant differences in the distribution of the angular core sizes of the strong and weak variables. This is true of angular core sizes not only that measured at 15 GHz where the variability is observed, but also at other frequencies (i.e., 2-24 GHz). This can be explained if the jets have a frequency-core size relationship, e.g., following the conical jet model (Ghisellini & Tavecchio 2009). The fact that we found that fast variable sources have significantly smaller angular core sizes, compared to the slow variable sources, can be explained by the light travel time argument. We are investigating the relationship between variability amplitudes/timescales with linear core sizes, for sources with redshift information (Hsu et al., in prep).

Acknowledgements

This research has made use of data from the OVRO 40-m monitoring program (Richards, J. L. et al. 2011, ApJS, 194, 29), supported by private funding from the California Institute of Technology and the Max Planck Institute for Radio Astronomy, and by NASA grants NNX08AW31G, NNX11A043G, and NNX14AQ89G and NSF grants AST-0808050 and AST- 1109911.

References

Alexander K. D. et al., 2020, *Space Sci. Rev.*, 216, 81
Ghisellini G. and Tavecchio F., 2009, *MNRAS*, 397, 985
Healey S. E. et al., 2008, *ApJS*, 175, 97
Hovatta T. et al., 2008, *A&A*, 485, 51
Hughes P. A., Aller H. D. and Aller M. F., 1985, *ApJ*, 298, 301
Koryukova T. A. et al., 2022, *MNRAS*, 515, 1736
Kovalev Y. Y. et al., 2005, *AJ*, 130, 2473
Kudryavtseva N. A. et al., 2011, *A&A*, 526, A51
Lin D. N. C. and Shields G. A., 1986, *ApJ*, 305, 28
Pushkarev A. B. and Kovalev Y. Y., 2015, *MNRAS*, 452, 4274
Richards J. L. et al., 2011, *ApJS*, 194, 29

The Multimessenger Chakra of Blazar Jets
Proceedings IAU Symposium No. 375, 2023
I. Liodakis, M. F. Aller, H. Krawczynski, A. Lähteenmäki & T. J. Pearson, eds.
doi:10.1017/S1743921323000650

Understanding the interplay between jets and ISM for winged radio galaxies

Gourab Giri and Bhargav Vaidya

Department of Astronomy, Astrophysics and Space Engineering, Indian Institute of Technology Indore, Simrol 453552, Madhya Pradesh, India
email: gourab@iiti.ac.in, bvaidya@iiti.ac.in

Abstract. In this work, we investigate the formation and early evolution phase of X-shaped radio galaxies using the Back-flow model. We show how the X-like winged morphology evolves over time in a tri-axial ambient medium, naturally. At this early stage of formation, we demonstrated that both the pair of jet lobes are actively pushing the ambient material out of their path of propagation, forming (X-ray) cavities that are surrounded by a shocked shell (X-ray bright rims) of swept materials. We also noticed how turbulent the wing is in comparison to the active lobe, generating sites of random shocks, indicating that the wings are not passively evolving structures. This study demonstrated that the ambiguous morphology observed in jets is also imprinted over the ambient medium, providing an alternative perspective in understanding the underlying physical process causing such ambiguities. Finally, we indicate that shearing instabilities cause mixing of ambient material at the shearing interface.

Keywords. Galaxies: jets, galaxies: active, Magnetohydrodynamics (MHD), Methods: numerical

1. Introduction

Extended radio galaxies show the presence of jets that emanate from the central Active Galactic Nuclei (AGN) and travel over a distance that can range from several hundreds of kilo-parsecs to a few mega-parsecs. As they travel through the heavy ambient medium (galactic, group, or cluster), they push aside the material that is circumjacent to them, which results in the formation of cavity regions. When studied using X-ray telescopes, an X-ray dip region (the cavities) is found to be surrounded by a bright X-ray envelope, also known as the rims. This phenomenon is caused by the expanding cocoon's shock-wave that has been generated by the backflow of materials of the inflating powerful (FR-II like) jets. Because of this, the jet is able to deposit a significant quantity of energy into the surroundings through the processes of heating, work done, shock waves, and turbulence (Cielo et al. 2018). The flow axis of bidirectional jets is typically predetermined by the spin axis of the supermassive blackhole (SMBH). As a result of this, it is not surprising to come across a pair of cavities that correspond with the flow of the straight, bidirectional jet. Recent findings with deep exposure maps, on the other hand, have shown X-ray cavities that have a high degree of complexity. For instance, Ubertoshi et al. (2022) reported two pairs of cavities that were equally spaced and positioned at a right angle to one another (see also Bogdan et al. 2014). A more complicated illustration can be seen in the Perseus Galaxy Cluster, which reveals off-axis cavities, rims, and intricate pressure waves (see also Ubertoshi et al. (2022) for RBS 797 galaxy cluster). An in-depth investigation has revealed that in each of the instances, the jets have most certainly deviated from the straight course that was anticipated for them, either as a result of jet precession or jet re-orientation into a different direction. These are some intriguing cases, and only a few of these sources have been thoroughly investigated. In general, it is apparent that the ambiguities noticed in the jet morphology also have an impact on the surrounding medium, which are highly intertwined. In this study, we will discuss the origin and evolution of a sub-class of unusual winged radio galaxies, and how they interact with its circumjacent medium. These galaxies exhibit two sets of jet lobes that are positioned at an angle to each other, raising X-like morphology (also known as X-shaped radio galaxies; XRGs).

The XRGs have a pair of lobes known as active lobes, which are mainly identified by the jet hotspots or by indications of active jet activity. The other pair of jet lobes has been identified as the secondary lobe, which was previously thought to evolve passively. Recent observations and simulation works, however, have provided evidence to the contrary. Due to their complex morphology and intrinsic properties, the formation process of such galaxies is still being debated. The mechanism of their formation can be broadly classified into two types (Gopal-Krishna et al. 2012): (a) back-flow model (in which the asymmetric ambient medium curves the back-flowing material in a different direction, producing an X-like structure) and (b) spin flip model (in which the initial jet ejection axis of the central SMBH is forced to change direction due to a BH-BH merger). We will use the back-flow model to generate the XRGs numerically, as it can naturally explain several key characteristics observed in these galaxies (Capetti et al. 2002). The primary goal of this work would be to demonstrate the formation and evolution of such galaxies based on the diverted back-flow scenario, as well as to demonstrate whether or not these RGs exhibit any signs of enhanced interaction with their surrounding ambient medium.

2. Simulation Setup

In order to perform the numerical simulation, we used the relativistic magneto-hydrodynamic (RMHD) module of PLUTO code (Mignone et al. 2007).
We started with an ellipsoidal (tri-axial) galaxy by employing the King's density profile (Cavaliere & Fusco-Femiano 1976), which is defined (in the Cartesian system) as

$$\rho = \frac{\rho_0}{(1 + x^2/a^2 + y^2/b^2 + z^2/c^2)^{\frac{3}{4}}} \tag{1}$$

where ρ_0 is the density at the centre of the galaxy (set to 1 amu/cc) and a, b and c are the effective core radius (4/3, 8/3 and 1 kpc, respectively) which make the density profile tri-axial. The galaxy has an initial isothermal atmosphere, whose static equilibrium is maintained by a gravity profile determined using the hydro-static equilibrium equation. We have introduced a jet nozzle at the centre of this distribution (i.e., at the galaxy centre) that will continuously inject bi-directional relativistic jets (radius 200 pc) having Lorentz factor $\Gamma = 5$ and which is under-densed as $10^{-6}\rho_0$. A B-field of toroidal nature is injected along with the jet which is defined as $B_x = -B_t r sin\vartheta$ and $B_y = B_t r cos\vartheta$, where r and ϑ are the polar coordinates in the perpendicular plane to the jet ejection axis, and B_x, B_y are the components of the field defined in that plane. The value of B_t is constant and is set to 100 μG. The parameter values are adopted from the study of Rossi et al. (2017) for the initial configuration, which have broader observational relevance. We simulate a 3D Cartesian domain having size $(16 \times 32 \times 16)$ kpc, distributed in $192 \times 384 \times 192$ grids, where y-axis is the jet propagation axis.

3. Jet-environment interaction & wing formation

At the initial time of jet propagation, the jet struggles to pierce the dense ambient medium, forming an ellipsoidal cocoon in which the jet cannot be identified individually. But, after a Mega-year of evolution, it extends beyond the core of the galaxy and begins

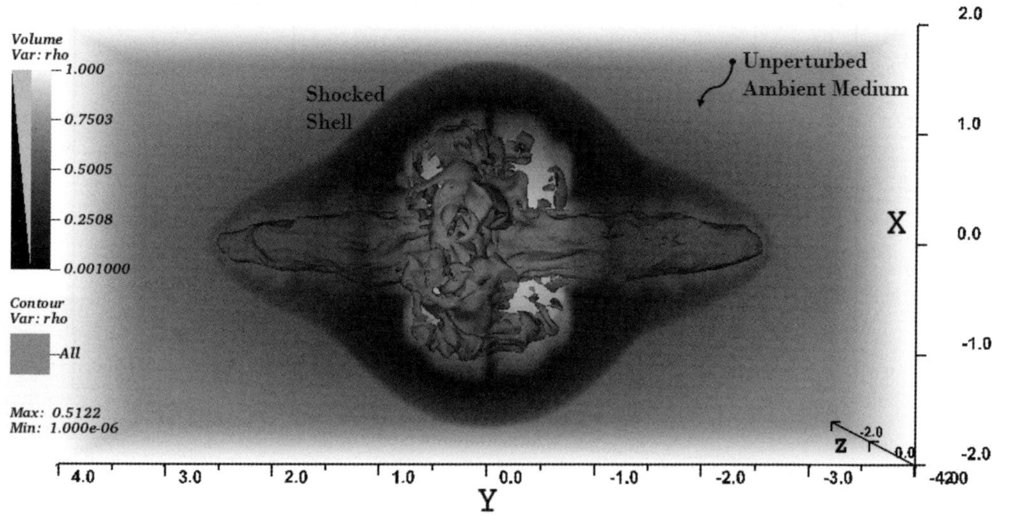

Figure 1. Volume rendered image of the simulated 3D density distribution, showing the jet-ambient medium configuration, indicating their highly interactive nature. The jet material here is represented with gray colored contour, which is overlaid with the ambient material indicated by the magenta color bar. Here, unit density is 1 amu/cc and unit length is 4 kpc.

to propagate rapidly. As the jet still remains underdense, it continues to generate a steady back-flow of matter from the unstable jet head. The asymmetric ambient medium further exerts additional force on the plasma material as it flows back towards the centre. This is the pressure gradient force that is higher along the minor axis of the galaxy. As a result of this force, the back-flowing plasma material diverts in the lateral direction, forming a structure that looks like a wing. We ran the simulation up to 4 Myr, just to investigate the early formation phase of such galaxies. At this stage, the active jet lobe still drills through the dense galactic medium with little decollimation. Therefore, the formed structure is expected to expand as it is for a significant course of it's evolution, which illustrates how XRGs with sizes of $\sim$100 kpc may have grown based on the Back-flow model. At this stage, the formed cocoon is still over-pressured and thus capable of removing the matter surrounding it (i.e., the ambient gas).

This is illustrated in Fig. 1, which is a volume rendered image of the density distribution of the jet-ambient system at time 3.9 Myr. It can be seen that the active jet is drilling through along the y-axis, concurrently with the construction of a wing structure, both of which are pushing the surrounding material away and producing a shell of over-density region. When observed with X-ray imaging telescopes with typical band ranges of 1-10 kev, we will see a cavity region and a bright rim (2D representation of the 3D shell structure) surrounding the cavities. It would be interesting to note that the ambiguities that are linked with the jet are produced by an underlying physical process, and the size and shape of the cavity region is expected to depend on these ambiguities. This can further be investigated in understanding the difference between various mechanisms that produces jet anomalies. In this regard, Fig. 2 shows the amount of volume the cocoon occupies within our simulated box at 3.9 Myr, as well as a comparison to the volume of an ellipsoid with a major axis of 16 kpc. This indicates how much fraction of the triaxial galaxy is disturbed by the newly-formed XRG. The small scale dips observed in the curve are due to the loss of cocoon volume caused by the mixing of jet-ambient material at the shear surfaces (due to shearing instabilities).

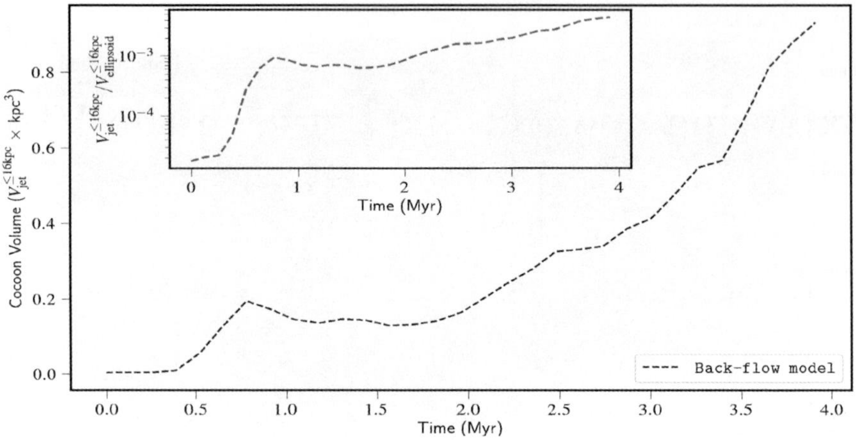

Figure 2. Evolution of the cocoon volume (in physical unit) and its comparison with the volume of an ellipsoid having a major axis of 16 kpc. This will give us an idea of how much of the ambient medium is affected by the inflated jet.

Lastly, we would like to highlight the turbulent wing structure that has been nicely captured in the volume rendered image in Fig. 1. The wing appeared more turbulent than the active lobe, producing sites of random shocks that in turn play a significant role in energizing the cooling particles, affecting it's emission signature. This has also been studied in Giri et al. (2022), using the evolution of passive particles.

4. Summary

In this work, we first showcase how a back-flow model can naturally form an X-like winged structure in a tri-axial medium. The primary motivation here was to demonstrate the enhanced interaction observed between such an unusual jetted system and its immediate ambient medium. The interaction caused by the expanding over-pressured cocoon pushes the surrounding matter aside, resulting in a shell-like shocked structure. When observed using imaging X-ray telescopes, such systems produce an X-ray cavity region at the center, surrounded by a bright rim of enhanced emission, which is a 2D representation of the 3D shocked shell shown here. We also demonstrate how turbulent the wing is, generating random shock sites, indicating that wings are actively evolving structures, at least in this early formation phase. Turbulence and shock sites observed in the wing play an important role in re-energizing the cooled particles, influencing their emission signature. Finally, we indicate that shearing instabilities cause significant mixing of jet material with the ambient material on the shear surface.

References

Bogdan, A, van Weeren, R, Kraft, R, et al. 2014, ApJL, 782, 2
Capetti, A., Zamfir, S., Rossi, P., et al. 2002, A&A, 394, 39
Cielo, S., Babul, A., Antonuccio-Delogu, V., et al. 2018, A&A, 617, A58
Cavaliere, A. & Fusco-Femiano, R. 1976, A&A, 500, 95
Giri, G., Vaidya, B., Rossi, P., et al. 2022, A&A, DOI:10.1051/0004-6361/202142546
Gopal-Krishna, Biermann P. L., Gergely L. Á., Wiita P. J., 2012, Research in Astronomy and Astrophysics, 12, 127
Mignone, A., Bodo, G., Massaglia, S., et al. 2007, ApJS, 170, 228
Rossi, P., Bodo, G., Capetti, A., & Massaglia, S. 2017, A&A, 606, A57
Ubertosi, F., Gitti, M., Brighenti, F., et al. 2022, ApJ [10.48550/arXiv.2212.10581]

The Multimessenger Chakra of Blazar Jets
Proceedings IAU Symposium No. 375, 2023
I. Liodakis, M. F. Aller, H. Krawczynski, A. Lähteenmäki & T. J. Pearson, eds.
doi:10.1017/S1743921323000662

Exploring connections between the VLBI and optical morphology of AGNs and their host galaxies

David Fernández Gil, Jeffrey A. Hodgson and Benjamin L'Huillier

Sejong University, Dept. of Physics and Astronomy, Seoul, South Korea

Abstract. We analyse VLBI and optical images of AGNs and their host galaxies and look for statistical correlations between the shape and orientation of the galaxy and the direction of the jet. We utilise the Astrogeo catalogue, which has over 9000 VLBI sources, many of those with a clear core-jet like structure that allows for the jet position angle to be reliably determined. We then use the VLBI source positions to search for optical counterparts within various optical surveys. In order to parameterise the orientation and shape of the host galaxy, we fitted a Gaussian elliptical model to the optical image, taking the PSF into account. We check our own shape parameters from this fit against the ones provided by the optical surveys. As of yet, no clear correlation between the galaxy morphology and the jet direction is seen.

Keywords. AGN jets | Host galaxies | Optical counterparts | VLBI | Astrogeo | SDSS | Jet direction | Galaxy morphology

1. Introduction

Active Galactic Nuclei (AGN) are compact central regions of galaxies that can produce relativistic jets. These jets, when pointed towards Earth, are referred to as blazars and can be detected via VLBI. AGN play an active role in star-formation feedback and are useful for studying distant galaxies due to their high brightness (Urry and Padovani 1995).

The study aims to investigate the correlation between small-scale jet properties and large-scale host galaxy characteristics, specifically examining the correlation between jet orientation and host galaxy optical shape. The expectation is that the jets may be perpendicular to the plane of the host galaxy; however, factors such as interaction with the environment, energy changes, projection and relativistic effects can complicate this relationship. The goal is to use a large sample to determine if there are any statistically significant generalities in the orientation of the jets relative to their host galaxies.

2. Observations

We utilize the Astrogeo VLBI catalogue, a geodetic catalogue, to gather data from over 9000 VLBI sources. The catalogue includes the coordinates of the sources and the jet position angle (PA, Plavin et al. 2022). The optical counterpart and shape parameters of the host galaxies are obtained from several optical surveys, indicated in Table 1.

3. Methods

We perform a cross-match by using the VLBI coordinates as a reference and searching for the closest optical source within 5 arcseconds. Shape parameters are obtained from published optical surveys (refer to Table 1). Aditionally, we download the images and

Table 1. Information about the optical surveys and their shape parameters.

Survey Name	Median Seeing	Band	Good fit definition	Reference
Sloan Digital Sky Survey (SDSS)	1.43"	r	type 'GALAXY'	Abdurro'uf et al. (2022)
Dark Energy Spectroscopic Instrument (DESI)	1.2"	g	type 'Sérsic'	Dey et al. (2019)
Kilo Degree Survey Golden Catalogue (KiDS)	0.7-0.9"	g	b > 2px, stellar index < 0.5	de Jong et al. (2013)
Skymapper	2.6"	g	b > 2px, stellar index < 0.5	Murphy et al. (2009)

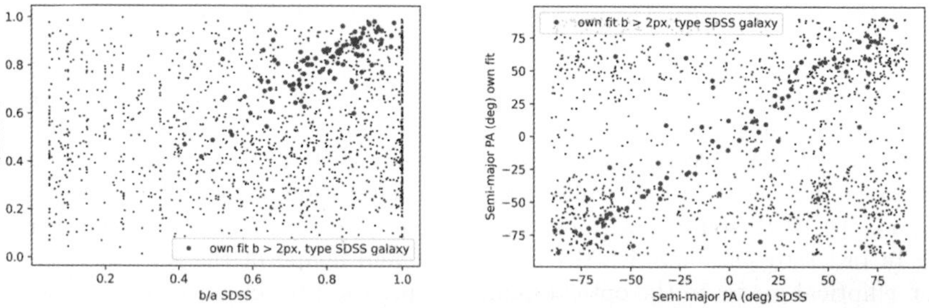

(a) Axis ratio.

(b) Semi-major axis position angle.

Figure 1. Comparison between our Gaussian fits and SDSS shape parameters. In red are the cases where the semi-minor axis of our fit is greater than 2 pixels and SDSS classifies the fit as 'Galaxy' type (resolved).

PSF of the optical counterparts in SDSS and forward-model their brightness distribution performing our own simple elliptical Gaussian fit (Finner et al. 2023). This provides the PA of the major and minor axis of the galaxy and the axis ratio. The VLBI jet's PA is then compared to the minor axis PA of the fitted ellipse, as this axis represents the perpendicular direction to the projected galaxy image.

4. Results

We first compare the results of our Gaussian fit with the parameters provided by the SDSS collaboration (Abdurro'uf et al. 2022). We define a good fit as when the semi-minor axis is greater than 2 pixels. A good fit from the SDSS is defined as when the catalogue classifies the source as type 'GALAXY' (resolved). In Figs. 1a and 1b we plot all fits (blue) and good fits (red) from both. We can see that the good fits are correlated.

We then include optical shape data from the other optical surveys that are listed in Table 1. For sources that are in both SDSS and DESI, we also compare our Gaussian fits with their shape parameters. A good fit in DESI is defined as when the shape model of the source is classified as 'Sérsic'. As can be seen in Figs. 2a and 2b, the good fits are also correlated. The horizontal lines of blue dots just show many cases where DESI could not resolve the shape, so it just assigned a PA of zero and axis ratio of one.

We perform a similar analysis using KiDS and Skymapper and find that for the good fit cases the results are correlated (see Table 1 for the definition of a good fit in each survey).

We find that the simple Gaussian fit performs well. However, there are some outliers. We individually inspect some of these cases and find that the outliers in the corners of Figure 1b correspond with cases where the 2-spin symmetry made the position angle of the semi-major axis jump from +90 to -90 even if the difference is slight. The other

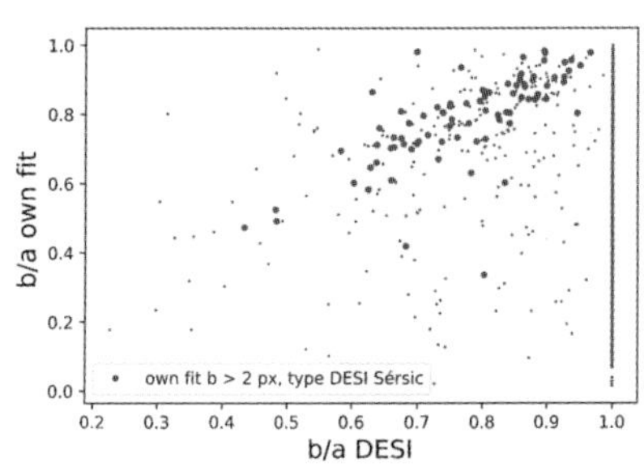

(a) Axis ratio.

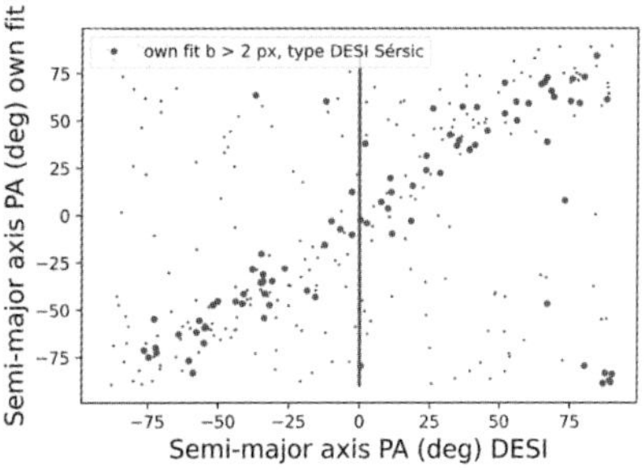

(b) Semi-major axis position angle.

Figure 2. Comparison between our Gaussian fits and DESI shape parameters. In red are the cases where the semi-minor axis of our fit is greater than 2 pixels and DESI classifies the fit as 'Sérsic' type (resolved). The horizontal lines of blue dots just show many cases where DESI could not resolve the shape, so it automatically assigned a PA of zero and axis ratio of one.

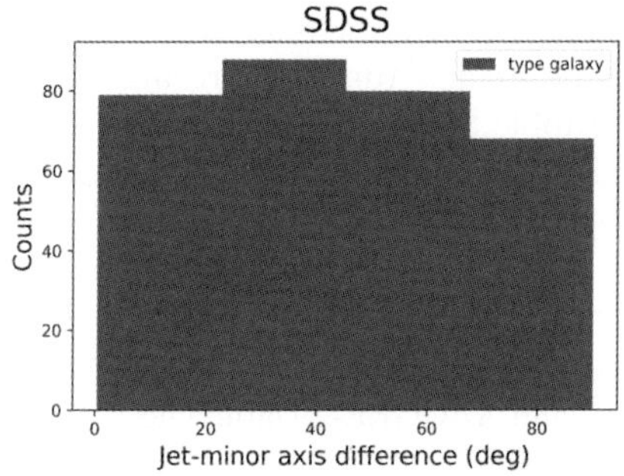

(a) SDSS.

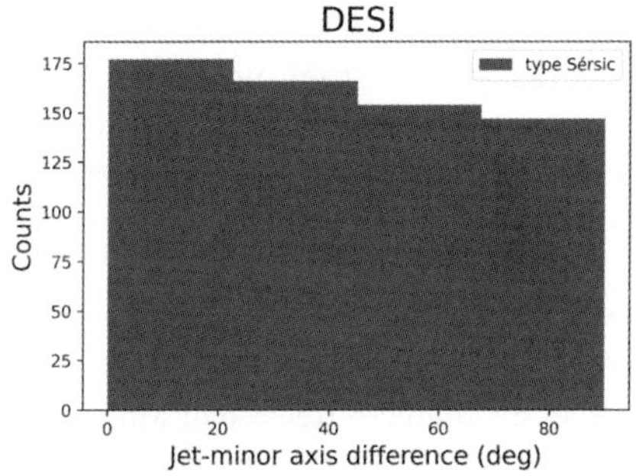

(b) DESI.

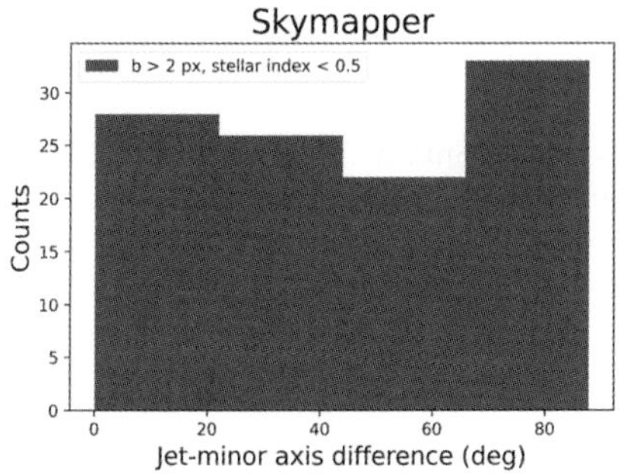

(c) Skymapper.

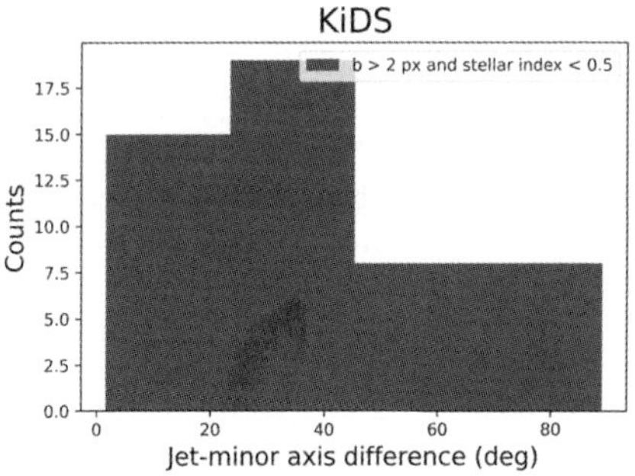

(d) KiDS.

Figure 3. Angle between the jet of the AGN and the semi-minor axis of the fitted shape of the host galaxies by the different optical catalogues, only including the cases where the fit was good, according to our criteria in Table 1.

outliers are QSO objects which, although marked as resolved by the catalogues, are actually not well resolved and could not be fit correctly.

Figs. 3a, 3b, 3c and 3d contain histograms with the angle between the jet and the semi-minor axis of the galaxies for each database, only using the 'good fit' cases as defined previously and in Table 1. In Figure 4, we plot a combined histogram with all surveys together, using the mean value of the angle difference if the source is present in several surveys.

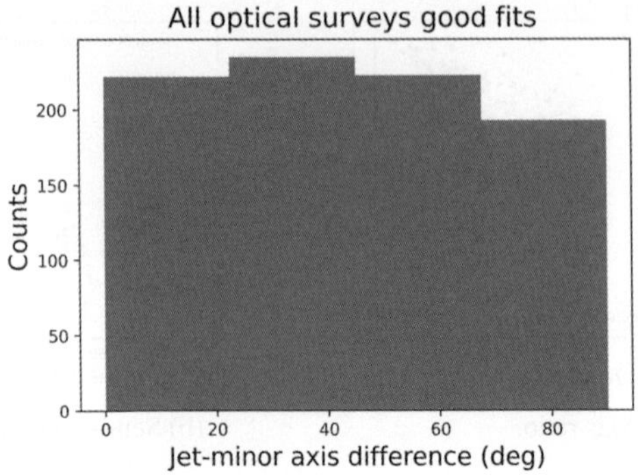

Figure 4. Combined histogram from the ones in Figs. 3a, 3b, 3c and 3d.

5. Discussion and conclusions

Overall, no clear angle preference for the jet is found. However, it is interesting that there may be hints of a signal in the highest resolution data (e.g. DESI and KiDS). Nevertheless at this stage no strong claims can be made. Adding more high resolution optical data (e.g. Dark Energy Survey and/or Hubble Space Telescope) may ultimately resolve the question. In any case, even if a signal is not detected, this still has important implications for fields as diverse as stellar and extra-galactic astrophysics.

References

Abdurro'uf and 340 colleagues 2022. The Seventeenth Data Release of the Sloan Digital Sky Surveys: Complete Release of MaNGA, MaStar, and APOGEE-2 Data. The Astrophysical Journal Supplement Series 259. doi:10.3847/1538-4365/ac4414

Dey, A. and 160 colleagues 2019. Overview of the DESI Legacy Imaging Surveys. The Astronomical Journal 157. doi:10.3847/1538-3881/ab089d

Finner, K. and 7 colleagues 2023. Near-IR Weak-Lensing (NIRWL) Measurements in the CANDELS Fields I: Point-Spread Function Modeling and Systematics. arXiv e-prints. doi:10.48550/arXiv.2301.07725

de Jong, J. T. A. and 58 colleagues 2013. The Kilo-Degree Survey. The Messenger 154, 44–46.

Murphy, S. and 6 colleagues 2009. SkyMapper and the Southern Sky Survey: A Valuable Resource for Stellar Astrophysics. The Eighth Pacific Rim Conference on Stellar Astrophysics: A Tribute to Kam-Ching Leung 404, 356. doi:10.48550/arXiv.0806.1770

Plavin, A. V., Kovalev, Y. Y., Pushkarev, A. B. 2022. Direction of Parsec-scale Jets for 9220 Active Galactic Nuclei. The Astrophysical Journal Supplement Series 260. doi:10.3847/1538-4365/ac6352

Urry, C. M., Padovani, P. 1995. Unified Schemes for Radio-Loud Active Galactic Nuclei. Publications of the Astronomical Society of the Pacific 107, 803. doi:10.1086/133630

The Multimessenger Chakra of Blazar Jets
Proceedings IAU Symposium No. 375, 2023
I. Liodakis, M. F. Aller, H. Krawczynski, A. Lähteenmäki & T. J. Pearson, eds.
doi:10.1017/S1743921323000820

RAD@home discovery of a one-sided radio jet hitting the companion galaxy

Ananda Hota[1,2], **Pratik Dabhade**[3,2] **and Sravani Vaddi**[4,2]

[1]UM-DAE Centre for Excellence in Basic Sciences, University of Mumbai, Mumbai-400098, India

[2]RAD@home Astronomy Collaboratory, Kharghar, Navi Mumbai, 410210, India

[3]Instituto de Astrofísica de Canarias, Calle Vía Láctea, s/n, E-38205, La Laguna, Tenerife, Spain

[4]Arecibo Observatory, NAIC, HC3 Box 53995, Arecibo, Puerto Rico, PR 00612, USA

Abstract. Minkowski's Object and 'Death Star galaxy' are two of the famous cases of rare instances when a radio jet has been observed to directly hit a neighbouring galaxy. RAD12, the RAD@home citizen science discovery with GMRT being presented here, is not only a new system being added to nearly half a dozen rare cases known so far but also the first case where the neighbouring galaxy is not a minor/dwarf companion but a galaxy bigger than the host of the radio jet. Additionally, the jet appears to be one-sided and the jet after interaction completely stops and forms a bubble inflating laterally which is unlike previous cases of minor deviation or loss of collimation. Since the nature of radio jet-ISM coupling is poorly understood so far, more discovery of objects like RAD12 will be important to the understanding of galaxy evolution through merger and AGN feedback.

Keywords. galaxies: active, galaxies: jets, galaxies: evolution, galaxies: interactions, radio continuum: galaxies

1. Introduction

Galaxy mergers and Active Galactic Nuclei(AGN) feedback have been the most favoured models to explain the observed properties of supermassive black holes and galaxies, namely the M-σ relation, the colour-magnitude diagram, the galaxy luminosity function, etc. (Di Matteo, Springel & Hernquist 2005; Heckman & Best 2014). Even after two decades of the proposed model, we are still in search of a "smoking gun" evidence showing the age of the AGN-driven outflow (cause) to precede the age of cessation of star formation (effect) in any sample of merger-remnant early-type galaxies (Hota et al. 2016). Furthermore, how the wind/jet fluid is coupled with the cold gas of the merger which may lead to momentarily positive or eventually negative feedback on star formation, is poorly understood. Here, we briefly report one extraordinary case, named RAD12, where radio jet outflow is seen interacting with the neighbouring galaxy in a possible major dry merger of two early-type galaxies. This was initially discovered by the RAD@home citizen science research collaboratory (Hota et al. 2014) and then followed up with the upgraded Giant Meterwave Radio Telescope (GMRT).

The basic methodology of citizen science research by the RAD@home Collaboratory† using GMRT telescope has been described in Hota et al. (2014, 2016). Briefly, it can be stated as a nationwide Inter-University collaboratory of trained citizen scientists,

† `https://www.radathomeindia.org`

professional astronomers and technical/administrative facilitators. It follows a hybrid method of both online and in-person training of citizens/students who have had basic University-level science education. The training is focused on ultraviolet(UV)-Optical-infrared(IR)-radio image analysis, through red-green-blue(RGB)-contour image overlays, by a large number of citizens/students who undergo training and report the discovery of interesting objects continuously. The platforms for interaction for growing the skill of RGB image interpretation are freely available Internet services, Facebook and Google. Citizens/students are trained in an organised way after a basic online selection process. Selected citizens who are trained extensively in a host institute for effectively one week are called e-astronomers. On the other hand, who are trained purely online or during one-day events with large gatherings of 50-200 members are called i-astronomers. One such one-day event was RAD@home Citizen Science Research Workshop organised during a science exhibition named *Vigyan Samagam* which was creating public awareness about the Square Kilometre Array (SKA). The week-long extensive training programmes are called *RAD@home Discovery Camps* which have been hosted by multiple national research institutes, research-cum-education institutes for undergraduate science and technology students, planetariums and science popularisation organisations of the Government of India. After nearly ten years of successful demonstration international expansion of the Collaboratory has been kick-started by the education and outreach programme of the IAU symposium 375 (see Rajoria et al. in this Proceeding). During such training and later on from their respective homes, trained citizen scientists discover unusual radio sources which are then followed up with the GMRT through a GMRT Time Allocation Committee-approved project titled GMRT Observation of Objects Discovered by RAD@home Astronomy Collaboratory (GOOD-RAC). One such unusual object (RAD12) followed up with the GMRT is being presented in this paper focusing on jet-galaxy interaction. The general scientific motivation of the collaboratory is to discover spiral-host radio galaxies like Speca (Hota et al. 2011) and cases of AGN-jet feedback in the context of galaxy mergers like 'cosmic leaf blower galaxy' NGC 3801 (Hota et al. 2012).

2. Discovery of RAD12

Radio-Optical-Radio Red-Green-Blue or ROR-RGB image overlays were the basic methods for trained citizen scientists to identify the host galaxy of a radio source. In the initial years, NASA SkyView was used but later we developed an RGB-maker web-tool (see Kumar et al. in this Proceeding). Figure. 1, presents typical image outputs from the RGB-maker†. In such images, an FRII-type radio galaxy would have two pink-looking radio emission blobs on either side of a green-looking optical host galaxy. In the case of an FRI-type radio galaxy, the peak of the elongated radio emission should coincide with the optical host galaxy. However, a peculiar case was spotted where the pink-looking radio emission blob was located in between two optical galaxies, both of which are like ellipticals in our IR-optical-UV or IOU-RGB images. Further comparison of NVSS and FIRST contour images (Figure. 1, right panel) clarified that the eastern optical galaxy coincides with the radio core, evidently the host, but NVSS emission, sensitive to diffuse emission, did not extend beyond the faint tail of emission seen in FIRST, sensitive to compact features. The faint radio tail seems to stop abruptly before reaching the western optical galaxy. It was suspected to be something else but not a head-tail radio galaxy, possibly a jet-galaxy interaction (Hota et al. 2014).

Details of the follow-up observations with the upgraded GMRT and of the supplementary archival data from the MeerKAT telescope are available in Hota et al. (2022). A

† https://www.radathomeindia.org/rgbmaker

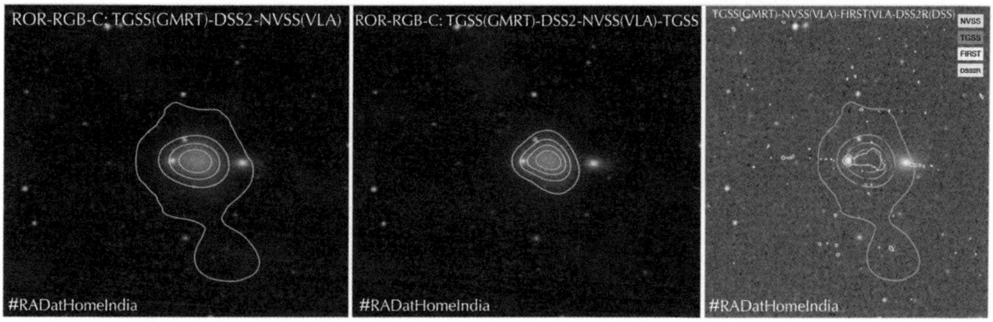

Figure 1. Radio-Optical-Radio image overlay of RAD12 in Red-Green-Blue channels and radio contours. This is a typical image output from RAD@home RGB-maker. The data on red, green, and blue channels are TIFR GMRT Sky Survey (TGSS ADR1), Digitized Sky Survey (DSS2 R) and NRAO VLA Sky Survey (NVSS) survey data. On the left panel, NVSS and on the middle TGSS radio contours are also plotted. The right panel plots the same optical data in greyscale along with TGSS (magenta), NVSS (cyan) and Faint Images of the Radio Sky at Twenty-Centimeters (FIRST; yellow) radio surveys. Note that the southern blob seen in NVSS is an unrelated radio source with optical and FIRST counterparts.

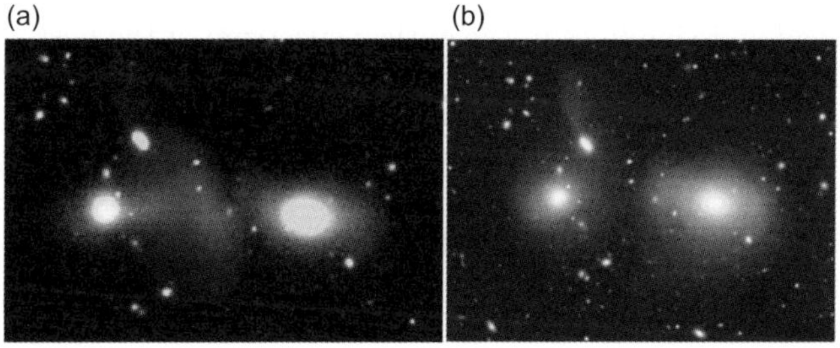

Figure 2. (a) Radio emission from GMRT in Red and MeerKAT in blue are overlaid on an optical image from Canada France Hawaii Telescope. Mushroom-shaped radio bubble seen in between two merging elliptical galaxies is clearly seen. (b) Deep optical image of RAD12 from legacy surveys, where the shells are prominently visible.

collage of radio emission (GMRT 323 MHz in red and MeerKAT L-band in blue) along with CFHT r-band optical data, in yellow, has been presented in Figure. 2a. GMRT detected the core-plume structure of the FIRST and resolved the NVSS blob to be a mushroom-shaped radio bubble. The tail of radio emission does not extend further to the west but shows a termination peak, supporting obstruction by the companion galaxy. The accompanying deep optical image (Figure. 2b) from the legacy‡ survey clearly shows multiple stellar shells in the western companion galaxy.

3. Discussion

In order to understand the overall radio spectral nature of RAD12, we have plotted the integrated flux densities of RAD12 with the available archival data ranging from $\sim$150 MHz to 5000 MHz as seen in Figure. 3. It is clear that the least square fit to the whole spectrum gives a spectral index of $\alpha \sim$ -0.59. This is similar to the injection spectral index of jets in radio galaxies suggesting either youth or re-acceleration of the old radio plasma (lacking jet or hotspot structure). The fainter outer regions (cap) of the mushroom bubble are steeper compared to the stem or central plume which is similar to FRII radio lobes. The radio bubble extends to nearly 137 kpc and unlike radio bubbles

‡ `https://www.legacysurvey.org/acknowledgment/`

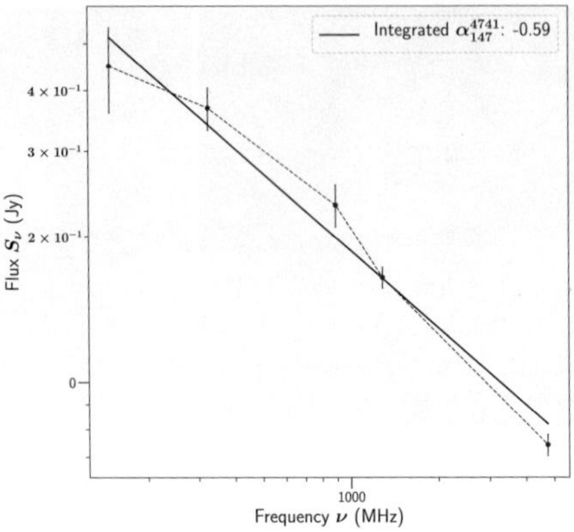

Telescope	Freq (MHz)	Core (mJy)	Total (mJy))
(1)	(2)	(3)	(4)
TGSS	147	-	449.58
GMRT	323	34.54	368.00
MeerKAT	1283	24.69	164.44
VLASS	3000	18.65	55.17
VLA	4741	18.60	44.51

Figure 3. Spectral fit for integrated flux densities for RAD12 from 147 MHz to 4741 MHz. The table presents flux densities of various components of RAD12 at various radio frequencies.

seen in nearby Seyfert galaxies (Hota & Saikia 2006). The spectral flattening and one-sidedness of the jet is unlikely due to the close alignment of the jet and line of sight as expected for a quasar.

The lack of broad emission lines in the optical spectrum of the host galaxy of RAD12, J004300.63–091346.3, as seen in Sloan Digital Sky Survey (SDSS), also does not favour the quasar nature of the AGN. Using the Wide-field Infrared Survey Explorer survey (*WISE*; Wright et al. 2010) we obtain the mid-infrared (mid-IR) magnitudes of RAD12. *WISE* mid-IR colours can be used to differentiate between quasars, star-forming, high and low excitation AGNs (see figure 3 of Dabhade et al. 2020). We adopt the scheme of Dabhade et al. (2020) and references therein to classify RAD12 as a low-excitation radio galaxy based on its colours (W1-W2 = 0.08 and W2-W3=1.19).

There are nearly half a dozen objects displaying radio jet hitting a neighbouring galaxy. They are Minkowski's Object near NGC541 (Croft et al. 2006), Death Star Galaxy or 3C321 (Evans et al. 2008), 3C34, 3C285, 3C441 and now RAD12 (Hota et al. 2022, and references therein). Hence, it is evident that even a dwarf galaxy can cause morphological distortions in the overall radio structure if it is in the direct path of a radio jet. Interaction with stellar shells has been proposed earlier to explain jet bending in Cen A (Gopal-Krishna & Saripalli 1984). Given the prominent stellar shells (Figure. 2b), we propose that the jet in RAD12 has likely been stopped by interstellar medium (ISM) trapped in the stellar shells. A preliminary look at the X-ray emission from the XMM-Newton data hints at the presence of a cavity, which needs further detailed investigation. From the limited radio-optical data available in archives and 323 MHz follow-up with the GMRT the newly discovered radio source RAD12 is demonstrated, in this very first imaging study, to be unique, justifying deep multi-wavelength follow-up by the community in radio polarisation, X-ray imaging and spectroscopy of ionised/molecular/atomic gas observations. It will be interesting if the gas kinematics can reveal how the radio bubble interacts with the ISM of the merging system, and hence will be very informative for AGN feedback studies.

Acknowledgement

We acknowledge financial support from the University Grants Commission (India), and International Astronomical Union for RAD@home team members due to which they could travel internationally and contribute to oral presentations, train international students for citizen science research and conduct outreach events. RAD@home thanks UM-DAE Centre for Excellence in Basic Sciences (Mumbai), Nehru Planetarium (NMML, Delhi) and International Centre for Theoretical Sciences - Tata Institute of Fundamental Research (Bengaluru (Code: ICTS/RADatICTS2018/05)) for hosting RAD@home Discovery Camps and contributing to the discovery of RAD12.

References

Croft, S., van Breugel, W., de Vries, W., et al. 2006, *ApJ*, 647, 1040.
Di Matteo, T., Springel, V., & Hernquist, L. 2005, *MNRAS*, 361, 776
Evans, D. A., Fong, W.-F., Hardcastle, M. J., et al. 2008, *ApJ*, 675, 1057.
Heckman, T. M. & Best, P. N. 2014, *ARAA*, 52, 589.
Dabhade, P., Mahato, M., Bagchi, J., et al. 2020, *A&A*, 642, A153.
Gopal-Krishna & Saripalli, L. 1984, *A&A*, 141, 61
Hota, A. & Saikia, D. J. 2006, *MNRAS*, 371, 945.
Hota, A., Sirothia, S. K., Ohyama, Y., et al. 2011, *MNRAS* (Letters), 417, L36.
Hota, A., Rey, S.-C., Kang, Y., et al. 2012, *MNRAS* (Letters), 422, L38.
Hota, A., Croston, J. H., Ohyama, Y., et al. 2014, ASI Conference Series, 13, 141.
Hota, A., Konar, C., Stalin, C. S., et al. 2016, *JA&A*, 37, 41.
Hota, A., Dabhade, P., Vaddi, S., et al. 2022, *MNRAS* (Letters), 517, L86.
Wright, E. L., Eisenhardt, P. R. M., Mainzer, A. K., et al. 2010, *AJ*, 140, 1868.

The Multimessenger Chakra of Blazar Jets
Proceedings IAU Symposium No. 375, 2023
I. Liodakis, M. F. Aller, H. Krawczynski, A. Lähteenmäki & T. J. Pearson, eds.
doi:10.1017/S1743921323000674

RAD@home RGB-maker web-tool for citizen science research in multi-wavelength study of AGNs with radio jets

Avinash Kumar[1], Ck. Avinash[1], Arundhati Purohit[1] and Ananda Hota[1,2]

[1]RAD@home Astronomy Collaboratory, Kharghar, Navi Mumbai-410210, India

[2]UM-DAE Centre for Excellence in Basic Sciences, University of Mumbai, Mumbai-400098, India

Abstract. The RAD@home RGB-maker Tool is a python-based web application that enables citizen science research through collaboration using open-source technology. The tool fetches FITS image data from NASA SkyView and generates false colour images in Red-Green-Blue channels with contour. This helps in the basic multi-wavelength understanding and characterization of extragalactic objects, and further analysis along with reporting of potential new discoveries in a uniform format. Students from Universities or science-graduate citizens gain skills in RGB-C image analysis and identify unique features in objects via either one-day online-weekend events or week-long in-person training. Trained citizen scientists in turn are part of rare discoveries such as the jet-galaxy interaction system RAD12. The tool has been successfully used by thousands of citizen scientists in India since its launch on 26th January 2021 and has demonstrated the potential during IAU symposium 375 to be part of citizen science efforts with international participants.

Keywords. Galaxy: evolution, radio continuum: galaxies, methods: data analysis

1. Introduction

Modern citizen science research in astronomy shares the processed data from telescopes with the public via websites, to simply classify the astronomical objects and report any unusual features. This often lacks one-to-one interaction between astronomers and citizen scientists. RAD@home (launched on 15th April 2013) in addition to providing data to the public, also trains them (science students or alumni of any University) through interactive sessions (online and in-person workshops) to discover interesting objects from large image files and then report them for follow-up observation or direct publication (e.g. Hota et al. 2016).

2. Summary

The RAD-RGB maker web-tool† uses data from surveys, like DSS2 IR-R-B (optical), WISE22-DSS2 R-GalexNUV (Infrared-Optical-Ultraviolet or IOU), and TGSS-DSS2R-NVSS (Radio-Optical-Radio or ROR) in Red-Green-Blue channels respectively with radio data in contours. Using these RGB-C images citizen scientists develop skills to identify the host galaxies of radio sources and further classification of radio emissions, which can be diffuse, compact or extended radio galaxies (FRI, FRII, DDRG, WAT, etc.).

† `https://radathomeindia.org/rgbmaker`

The tool also helps in the uniform reporting of unusual objects with additional multi-wavelength characterization from open databases and literature. Building on the pre-discovery reports of such unusual objects, the Collaboratory then submits proposals to suitable telescopes for follow-up observations. One recent example is the discovery of RAD12 where the radio jet from one galaxy is interacting and most plausibly bouncing back from the companion (Hota et al. 2022).

References

Hota, A., Konar, C., Stalin, C. S., et al. 2016, *JA&A*, 37, 41.
Hota, A., Dabhade, P., Vaddi, S., et al. 2022, *MNRAS* (Letters), 517, L86.

The Multimessenger Chakra of Blazar Jets
Proceedings IAU Symposium No. 375, 2023
I. Liodakis, M. F. Aller, H. Krawczynski, A. Lähteenmäki & T. J. Pearson, eds.
doi:10.1017/S1743921323000686

RAD@home Inter-University Collaboratory for citizen science in galaxy evolution with multi wavelength RGB images

Megha Rajoria[1], **Arundhati Purohit**[1] **and Ananda Hota**[1,2]

[1]RAD@home Astronomy Collaboratory, Kharghar, Navi Mumbai-410210, India

[2]UM-DAE Centre for Excellence in Basic Sciences, University of Mumbai, Mumbai-400098, India

Abstract. Unlike most web-based citizen science programmes, RAD@home citizen scientists are trained in an interactive manner, both online and in person. This Collaboratory model can alleviate various geo-political and socio-economic constraints on the growth of people in underdeveloped regions which is in synergy with mandates of the IAU Office of Astronomy for Development. Owing to its Inter-University collaboratory design, supported by over 30 research institutions, it can attempt to convert the 'Big Data Problem' into a 'Big Data Prospect' by direct training and involving citizens in discoveries from multi-telescope data. Using the RAD-RGB-maker web-tool, e-astronomers learn the basics of multi-wavelength (UV-optical-IR-radio) extragalactic astronomy and create preliminary discovery reports of unusual radio sources thus becoming Co-Investigator in follow-up observations and publications (e.g. Discovery of RAD12). International expansion of RAD@home has been initiated during the recent IAU symposium 375 through its Educational and Outreach programme.

Keywords. Galaxy: evolution, radio continuum: galaxies, methods: data analysis

1. Summary

Professional astronomers are sometimes overloaded with data and most students on the other hand are waiting for involvement in a project. Citizen science research through web interfaces is a promising solution. RAD@home citizen science research Collaboratory† combines the web interface with face-to-face interaction to solve the problems of both the scientists as well as students/citizens (e.g. Hota et al. 2016). Professional astronomers in our collaboratory train citizen scientists in camps covering various basic aspects of the subject as well as a broad overview. In the past 10 years, RAD@home has conducted nearly 50 such camps around the country, which has resulted in the training of thousands of citizens. In order to facilitate learning and discovery in a uniform and relatively easier way, we have developed a web-tool RGB-maker (Kumar et al. 2023 this IAU 375 Proceeding). This tool can be used for introduction to multi-wavelength (UV-optical-IR-radio) image analysis of galaxies. Then, they can be asked to report any faint peculiarly shaped radio emission for follow-up observation and/or publication. An example of the process is the discovery of RAD12, a one-sided radio jet hitting the merging companion galaxy and probably bouncing back, a laboratory for AGN-feedback (Hota et al. 2022). Various international organisations, including IAU‡, and several

† `https://radathomeindia.org/`
‡ `https://astro4edu.org/shaw-iau/sessions/citizen-science/`

dozen national institutions appreciate Collaboratory model and support for further its growth§¶.

References

Hota, A., Konar, C., Stalin, C. S., et al. 2016, *JA&A*, 37, 41.
Hota, A., Dabhade, P., Vaddi, S., et al. 2022, *MNRAS* (Letters), 517, L86.

§ `https://vigyansamagam.in/mega-science-projects/ska/`
¶ `http://nopr.niscpr.res.in/handle/123456789/61164/`

The Multimessenger Chakra of Blazar Jets
Proceedings IAU Symposium No. 375, 2023
I. Liodakis, M. F. Aller, H. Krawczynski, A. Lähteenmäki & T. J. Pearson, eds.
doi:10.1017/S1743921323001205

3D PIC Simulations for Relativistic Jets with a Toroidal Magnetic Field

Kenichi Nishikawa[1], Athina Meli[2,3], Christoph Köhn[4], Ioana Duţan[5], Yosuke Mizuno[6,7], Oleh Kobzar[8,9], Nicholas MacDonald[10], José L. Gómez[11] and Kouichi Hirotani[12]

[1]Alabama A&M University,

[2]North Carolina A&T State University,

[3]Institute Universite de Liege,

[4]Technical University of Denmark,

[5]Institute of Space Science,

[6]Shanghai Jiao Tong University,

[7]Goethe University,

[8]Astronomical Observatory of the Jagiellonian University,

[9]Cracow University of Technology,

[10]Max-Planck-Institut für Radioastronomie,

[11]Instituto de Astrofísica de Andalucía,

[12]Taiwan Institute of Astronomy and Astrophysics, Academia Sinica

Abstract. Particle-in-Cell simulations can provide a possible answer to an important key issue for astrophysical plasma jets; namely on how a toroidal magnetic field affects the evolution of pair and electron-ion jets. We show that Weibel, mushroom, and kinetic Kelvin-Helmhotz instabilities excited at the linear stage, generate a quasi-steady x-component of the electric field which accelerates and decelerates electrons. We observe significant differences in the structure of the strong electromagnetic fields that are driven by the kinetic instabilities with the pair jet. We find that the two different jet compositions ($e^{\pm}$ and e^{-} - i^{+}) generate different instability modes respectively. Moreover, the magnetic field in the non-linear stage generated by different instabilities is dissipated and reorganized into new topologies. A 3D magnetic field topology depiction indicates possible reconnection sites in the non-linear stage where the particles are significantly accelerated by the dissipation of the magnetic field associated to a possible reconnection manifestation.

Keywords. relativistic jets, kinetic instabilities, particle acceleration, magnetic reconnection

---

1. Introduction

Relativistic astrophysical jets are ubiquitous in astrophysical systems, and have been observationally associated with the activity of central black holes in Active Galactic Nuclei (AGN, e.g., EHT Collaboration 2019). It is also theorized that relativistic jets may occur in Gamma-ray Bursts (GRBs, e.g, Ruiz *et al.* 2018). The formation and powering of these astrophysical jets are highly complex phenomena involving relativistic plasmas and twisted magnetic fields which are organized in such a manner as to ultimately launch an outflow from a central compact source (e.g., McKinney & Uzdensky 2012).

Relativistic jets interact with the plasma environment of an astrophysical source and subsequently instabilities occur which are responsible for the acceleration of particles (e.g., Nishikawa *et al.* 2021). In some cases, e.g. for unmagnetized jets, previous computer simulations have shown that the Weibel instability (WI) excites relativistic shocks, which results in particle acceleration (e.g., Nishikawa *et al.* 2009, and references therein), Other instabilities such as the kinetic Kelvin-Helmhotz (kKHI) and the mushroom instability (MI), are driven by the velocity-shear at the boundary between the jet and the ambient medium, which were first observed in using a slab model for plasma sheets (e.g., Alves *et al.* 2012; Nishikawa *et al.* 2014). PIC simulation studies have been performed for the evolution of cylindrical jets with a helical magnetic-field topology (e.g., Alves *et al.* 2015; Nishikawa *et al.* 2021).

Moreover, using the so-called Harris model in a slab geometry, magnetic reconnection has commonly been studied with PIC simulations. It was observed to produce a significant particle acceleration (e.g., Sironi & Spitkovsky 2014; Guo *et al.* 2016).

Global 3D PIC modeling of relativistic jets allows for a self-consistent investigation of the complex kinetic processes occurring in the jet and the surrounding medium. These processes can reveal electron-scale short-wavelength instabilities, their saturation and associated phenomena. Such studies have been first performed for unmagnetized jets (Nishikawa *et al.* 2016a). PIC simulations of relativistic jets containing helical magnetic fields were, for the first time, presented by Nishikawa *et al.* (2016b). These initial studies addressed the early, linear growth of kinetic instabilities in the electron-ion and electron-positron jets. However, such simulations were limited by the size of the computational box.

1.1. *Importance of toroidal magnetic field*

To address the importance of toroidal magnetic fields for growing instabilities, the present study involves a much larger jet radius and longer simulation times than previous works (e.g., Nishikawa *et al.* 2016a; Nishikawa *et al.* 2016b), allowing for a non-linear evolution of the jets with a toroidal magnetic field aiming the following key questions: (i) How does a toroidal magnetic field affect the growth of kKHI, MI, and WI within the jet and in the jet-ambient plasma boundary? (ii) How do jets composed of electrons and positrons and jets composed of electrons and ions evolve in the presence of a large-scale toroidal magnetic field? (iii) How and where are particles accelerated in jets with different plasma compositions?

2. Simulation set-up

Our 3D PIC code is a modified version of the relativistic electromagnetic PIC code TRISTAN (Buneman 1993) with MPI-based parallelization (Niemiec *et al.* 2008; Nishikawa *et al.* 2009). The numerical grid is set to $(L_{\rm x}, L_{\rm y}, L_{\rm z}) = (1285\Delta, 789\Delta, 789\Delta)$ and is twice as long as that in our previous simulation studies (Nishikawa et al. 2016b). Here $\Delta = 1$ is the size of an individual grid cell. Open boundaries are used on the surfaces at $x/\Delta = 0$ and 1285, whilst periodic boundary conditions are implemented along the transverse directions y and z. Since the jet is located in the center of the simulation box far from the boundaries, the effect of periodic boundaries is negligible.

3. Simulation results

We present simulation results for $\mathrm{e}^{\pm}$ and e^{-} - i^{+} jets with a toroidal magnetic field, applying a new and improved jet injection scheme (for details, see Meli *et al.* 2023). We are in particular interested in the differences in the dynamical behavior of the jets of different plasma compositions and in the way these jets interact with the surrounding

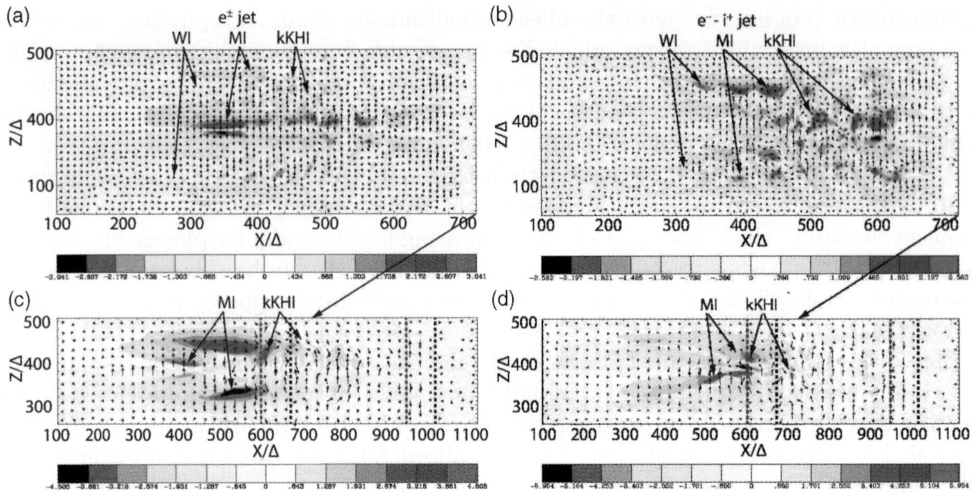

Figure 1. Color maps of the B_y magnetic field with arrows depicting the magnetic field components in the x - z plane, at $t = 900\,\omega_{pe}^{-1}$. Upper panels (a) and (b) show unmagnetized jets and lower panels (c) and (d) jets with a toroidal magnetic field. The maximum and minimum are (a): ±1.525, (b): ±4.073, (c): ±4.505 and (d): ±5.954. Adapted from Fig. 4 in Meli *et al.* (2023).

environment. In order to give an overview of how the toroidal magnetic field affects the evolution of the jet, in Fig. 1 we present simulation results for jets with a toroidal field (shown in the lower panels of each figure) and compare them to the results obtained for an unmagnetized jet.

Figure 1 shows the amplitude of B_y component with the arrows indicating the magnetic field components in the x - z plane. For the magnetized jets, we apply $B_0 = 0.5$ initially at the jet orifice while B_y is measured in the same units. In the presence of the toroidal magnetic field for the $e^{\pm}$ jet, we find a maximum value of $B_y = 4.505$, which means that B_0 is amplified by a factor of 9.01 over the initial value. It is a factor of 2.95 stronger than magnetic fields in the unmagnetized $e^{\pm}$ jet. The evident differences in the magnetic field structure in cases with and without the initial field, indicate a significant impact of the toroidal field on the development of the kinetic instabilities. The same is true for the e^- - i^+ jet, in which the magnetic field amplification is comparably stronger and the field amplitudes reach $B/B_0 \approx 10.6$, a factor of 1.3 stronger compared to the unmagnetized case. It should be noted that for unmagnetized e^- - p^+ jet the large mass ration contributes to a stronger growth of MI, this is the reason why we needed to reduce the mass ratio to control the growth of MI. We note that for the $e^{\pm}$ jet the magnetic field dissipates, i.e., becomes considerably weakened, at the jet region $x/\Delta \gtrsim 680$ and similarly for the e^- - i^+ jet the weakening occurs around $x/\Delta \gtrsim 700$. By comparing both magnetized jet species, at the non-linear stage close to $x/\Delta \gtrsim 950$, one discerns that for the electron-ion jet the B_y field almost dissipates and becomes disorganized (turbulent).

4. Summary

We have conducted extensive 3D PIC simulations to study the spatiotemporal evolution of magnetized relativistic electron-positron and electron-ion jets, examining their kinetic instabilities and the associated particle acceleration. We investigated the excited kinetic instabilities and the associated magnetic fields at the linear and non-linear stage, as they may occur in astrophysical relativistic jets. The dissipation of the magnetic fields was observed to generate electric fields that are sufficiently strong to further accelerate particles to Lorentz factors of up to around 35.

In this work we used a new jet injection scheme. We injected both $e^{\pm}$ and e^{-} - i^{+} jets, with a co-moving toroidal magnetic field while we used a top-hat jet density profile. The current was self-consistently carried by the jet particles, $\mathbf{J} = \nabla \times \mathbf{B}$. In order to sustain the toroidal magnetic field carried by the jet, the current was applied at the jet orifice and a motional electric field was applied in order to compensate the bending by the applied toroidal magnetic field. Both jets were initially moderately magnetized while the ambient medium remained unmagnetized. We have run the simulations sufficiently long in order to examine the non-linear effects of the jet evolution.

We found that the dominant excited modes of instabilities depend on the different jet compositions. Three different instabilities (WI, MI, kKHI) grow in similar timescales, however, depending on the plasma conditions (unmagnetized, different species) some instabilities grow faster and stronger. Particularly the MI grows faster and stronger with larger mass ratio (4 and 1836) and an excitation of MI or kKHI is weaker for the unmagnetized case. In general, stronger MI and kKHI grow in the jets with toroidal magnetic fields. The MI and kKHI are associated with a quasi-steady electric field ($E_{\rm x}$) for both jet species. These accompanied electric fields accelerate and decelerate electrons and positrons. Additionally we found that the electrons can be further accelerated by the development of twisted and turbulent magnetic fields which are generated by the dissipation of the toroidal magnetic fields possibly accompanying reconnection.

We have identified potential sites of magnetic reconnection in our simulations; however, an unambiguous determination in 3D is not trivial (see, Meli *et al.* 2023). The magnetic field structure of a reconnection site in 2D simulations consists of X and O shapes which can be recognized rather easily by the changes of the magnetic field direction and the position of null (very weak) magnetic fields in the 2D projections. The complex structures of 3D reconnections have been investigated in (e.g., Lazarian *et al.* 2020). In order to determine the reconnection locations analytically, we would need to investigate the eigenvalues of Jacobian matrix, which is beyond the scope of this work, for more details, see Cai *et al.* (2007).

For an $e^{\pm}$ jet, the MI is excited combined with a kKHI, while the produced quasi steady $E_{\rm x}$ modulates the jet particles. For the electron-ion jet, the jet electrons are pinched dominantly grown MI at the later linear stage. Further simulations will be important to decisively confirm possible supplemental acceleration mechanisms with varying simulation parameters such as, jet radius, magnetization factor, jet density profile, etc.

From the present extensive simulation studies we conclude, that a moderate initial magnetic field can change the development of the kinetic plasma instabilities, for different jet species, even if the instabilities significantly amplify the magnetic field, here by Lorentz factors reaching 50.

References

Alves, E. P., Grismayer, T., Martins, S. F., *et. al* 2012, *ApJ*, 746, L14

Alves, E. P., Grismayer, T., Fonseca, R. A., *et. al* 2015, *Phys. Rev. E*, 92, 021101

Buneman, O. 1993, in Computer Space Plasma Physics: Simulation Techniques and Software, Eds.: Matsumoto & Omura, Tokyo: Terra, p. 67

Cai, D. S., Nishikawa, K.-I., & Lembege, B. 2007, in Advanced Methods for Space Simulations, Eds.: H. Usui and Y. Omura, TERRAPUB, Tokyo, p. 145

EHT Collaboration 2019, *ApJL*, 875, L1

Guo, F., Li, H., Daughton, W., *et. al* 2016, *PhPl*, 23, 5708

Lazarian, A., Eyink, G. L., Jafari, A., *et. al* 2020, *Phys. Plasmas*, 27, 012305

McKinney, J. C. & Uzdensky, D. A. 2012, *MNRAS*, 419, 573

Meli, A., Nishikawa, K.-I., Köhn, C., *et. al* 2023, *MNRAS*, 519, 5410
Niemiec, J., Pohl, M., Stroman, T., & Nishikawa, K.-I. 2008, *ApJ*, 684, 1174
Nishikawa, K.-I., Niemiec, J., Medvedev, M., *et. al* 2009, *ApJ*, 698, L10
Nishikawa, K.-I., Hardee, P. E., Duţan, I., *et. al* 2014, *ApJ*, 793, 60
Nishikawa, K.-I., Frederiksen, J. T., Nordlund, Å., *et. al* 2016a, *ApJ*, 820, 94
Nishikawa, K.-I., Mizuno, Y., Niemiec, J., *et. al* 2016b, *Galaxies*, 4, 38
Nishikawa, K.-I., Mizuno, Y., Gómez, J. L., *et. al* 2020, *MNRAS*, 493, 2652
Nishikawa, K.-I., Duţan, I., Köhn, C., & Mizuno, Y., Living Reviews in Computational Astrophysics 2021, 7, 1
Ruiz, M., Shapiro, S. L., & Tsokaros, A. 2018, *Phys. Rev. D.*, 97, 021501
Sironi, L. & Spitkovsky, A. 2014, *ApJ*, 783, L21

The Multimessenger Chakra of Blazar Jets
Proceedings IAU Symposium No. 375, 2023
I. Liodakis, M. F. Aller, H. Krawczynski, A. Lähteenmäki & T. J. Pearson, eds.
doi:10.1017/S1743921323000637

Effect of Inverse Compton Cooling on Relativistic Particles Accelerated at Shear Boundary Layers in Relativistic Jets

Tej Chand and Markus Böttcher

Centre for Space Research, North-West University, Potchefstroom, 2520, South Africa
emails: chandtej11@gmail.com, Markus.Bottcher@nwu.ac.za

Abstract. Recent theoretical considerations and observational evidence evince the spine-sheath morphology of relativistic jets emitted from active galactic nuclei (AGNs) or gamma-ray bursts (GRBs). The resulting shear boundary layers (SBLs) are likely to be an avenue for particle acceleration in relativistic jets. The effect of radiation drag on radiating particles has yet to be addressed in most studies of particle acceleration at shear boundary layers, even though radiative cooling may considerably affect particle dynamics. By using particle-in-cell simulations, we study the effects of inverse Compton cooling on particle dynamics and emerging particle spectra.

Keywords. shear boundary layers, shear acceleration, radiation mechanism, radiation-drag

1. Introduction

The relativistic jets emanating from the central engine of active galactic nuclei (AGNs) or gamma-ray bursts (GRBs) interact with ambient plasma resulting in jet stratification in the radial direction. In this picture, the jet is considered to consist of a fast-moving inner spine surrounded by a slow-moving outer sheath (e.g., Ghisellini et al. 2005). The resulting shear boundary layers (SBLs) are promising sites for particle acceleration in relativistic jets. Owen et al. (1989) first suggested the existence of radial stratification of the relativistic jets with velocity shear to explain the VLA radio observations of the M87 jet. Shear acceleration can be defined as a second-order-Fermi-type acceleration mechanism and is expected in fast, velocity-shearing plasma flows. SBLs in relativistic jets are subject to plasma instabilities like the Weibel instability (Weibel 1959), and two-stream instability (Boyd et al. 2003), which gives rise to the self-generation of electric and magnetic fields (e.g., Liang et al. 2013; Chand et al. 2019). The self-generated magnetic field develops the turbulence in SBLs from initially unmagnetized plasma, which eventually leads to particle acceleration (e.g., Alves et al. 2012; Liang et al. 2013). The particle acceleration in shear boundary layers has been broadly studied using PiC simulations due to its obvious astrophysical significance. However, most of these studies have overlooked the possible effect of radiation drag on particle dynamics.

The two prime radiative processes occurring at the SBLs of relativistic jets are synchrotron emission and inverse Compton (IC) emission due to the interaction of the relativistic electrons with the synchrotron photons (synchrotron self-Compton) or an external seed photon field (external Compton). We consider an inverse-Compton cooling of the relativistic electrons accelerated at SBLs due to the blackbody photon field in the Thomson regime. In particular, we investigate the effect of IC cooling on the particle

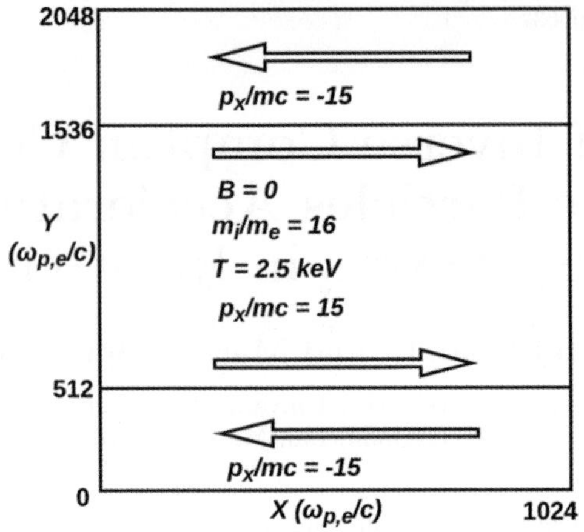

Figure 1. 2D simulation setup of the initially unmagnetized shear flow of the electron-ion plasma: the plasma consists of right-moving plasma (spine) in the central 50% of the Y-grid crammed between left-moving plasmas (sheath) occupying top 25% and bottom 25% of the Y-grid. Spine and sheath counterstream with equal and opposite x-momenta $p_x/mc = \pm 15$ in the ELF frame.

energy distribution at SBLs of relativistic jets. In section 2, we discuss our numerical model setup with the initial simulation parameters. In section 3, we describe a theoretical background of the IC cooling due a to blackbody target photon field and present the resulting particle spectra. We summarize our results in section 4.

2. Numerical Simulation Setup

We set up a 2D particle-in-cell (PiC) simulation of self-generation of B- and E- fields from initially unmagnetized plasma at SBLs in relativistic jets using the code TRISTAN-MP (Spitkovsky 2005). The dimension of the simulation boxes is 1024×2048 with shear boundary layers located at $Y = 512$ and $Y = 1536$ (see fig. 1). In the simulations, distances are measured in units of plasma skin depth (i.e., $\Delta x = c/\omega_{p,e}$), and time is measured in terms of inverse plasma frequency, ($\Delta \mathrm{t} = 1/\omega_{\mathrm{p,e}}$), where the plasma frequency is given by, $\omega_{\mathrm{p,e}} = \sqrt{4\pi \mathrm{ne}^2/\mathrm{m_e}}$ (n, e, and $\mathrm{m_e}$ indicate electron density, electron charge, and electron mass, respectively), and c is the speed of light. The numerical speed of light is considered as $\mathrm{c} = 0.45\Delta \mathrm{x}/\Delta \mathrm{t}$ to satisfy the Courant condition. We simulate 20 particles per cell with fiducial ion-to-electron mass ratio, $(\mathrm{m_i}/\mathrm{m_e}) = 16$. Our simulations are performed for pure electron-ion plasma. The upper and lower part of the simulation grid occupying 50 % of the box constitutes the sheath region, and the remaining central portion of the grid represents the spine region. The simulations are performed in equal Lorentz factor (ELF) frame of reference. The spine and sheath, with a uniform temperature of 2.5 keV, counterstream with an equal bulk Lorentz factor, $\Gamma = 15$.

3. Effect of Radiation Drag on Particle Spectra

The inverse Compton scattering of a single electron of Lorentz factor γ due to interaction with photons of energies $\varepsilon = \mathrm{h}\nu/\mathrm{m_e c^2}$ is described by the inverse Compton spectrum peaking at $\gamma^2\varepsilon$. We use a simple δ-function approximation with a characteristic photon frequency of $\mathrm{h}\nu = 2.7\mathrm{K_B T}$ ($\mathrm{K_B}$ is the Boltzmann's constant) for the target photon field to calculate the inverse Compton cooling term. The radiation cooling term for inverse

Compton scattering of relativistic electrons in the angle-integrated blackbody photon field in the Thomson regime and boosted to the spine frame is

$$\frac{d\gamma}{dt} = \Gamma_{rel}^2 \frac{\pi^4}{15} \sigma_T c K \gamma^2 \theta^4, \tag{1}$$

where $K = 8\pi/\lambda_C^3$, $\theta = K_B T/m_e c^2$, $\sigma_T = 8\pi e^4/3m_e^2 c^4$ is the Thomson cross section, and $\gamma = 1/\sqrt{1-\beta^2}$, with $\beta = v/c$ (v refers to the electron velocity). Here, λ_C, T, and e are Compton wavelength, radiation temperature, and electron charge, respectively. $\Gamma_{rel} = 2\Gamma^2 + 1$ is the relative bulk Lorentz factor between the spine and the sheath.

Considering the head-on collision between the electron and photons, the Compton cross section in the case of an angle-dependent target photon field can be approximated (Boettcher, Harris, & Krawczynski, 2012) as

$$\frac{d\sigma_C}{d\Omega_s d\varepsilon_s} = \frac{\pi r_e^2}{\gamma \varepsilon'} \left\{ y + \frac{1}{y} - \frac{2\varepsilon_s}{\gamma \varepsilon' y} + \left(\frac{\varepsilon_s}{\gamma \varepsilon' y} \right)^2 \right\} H\left(\varepsilon_s \, ; \, \frac{\varepsilon'}{2\gamma}, \, \frac{2\gamma \varepsilon'}{1 + 2\varepsilon'} \right), \tag{2}$$

where H is the Heaviside function that constrains the upper and lower limits of the scattered photon energies, and $y = 1 - \frac{\varepsilon_s}{\gamma}$. $\varepsilon' = \gamma\varepsilon(1 - \beta\mu)$ is the initial photon energy in the electron's rest frame. $\mu = \cos\psi$, where ψ is the angle between the direction of propagation of the electron and photon. From equation (2), the Compton emissivity in the spine frame can be written as

$$\mathrm{j}(\varepsilon, \gamma, \mu) = \frac{3 m_e c^3 \sigma_T \varepsilon_s \Gamma_{rel}^2 (1 - \beta\mu)}{8\gamma} \int_0^\infty \frac{1}{\varepsilon'} \left\{ y + \frac{1}{y} - \frac{2\varepsilon_s}{\gamma \varepsilon' y} + \left(\frac{\varepsilon_s}{\gamma \varepsilon' y} \right)^2 \right\} n_{ph}(\varepsilon, \theta)$$
$$H\left(\varepsilon_s ; \frac{\varepsilon'}{2\gamma}, \frac{2\gamma\varepsilon'}{1 + 2\varepsilon'} \right) d\varepsilon$$

With the δ-function approximation employed for the spectral calculation, the photon density is given by $n_{ph}(\varepsilon, \theta) = n_{ph}(\theta)\delta(\varepsilon - 2.7\theta)$. The photon density now simplifies to $n_{ph}(\theta) = 2.4K\theta^3$. After some simplifications, $\mathrm{j}(\varepsilon, \gamma, \mu)$ can be integrated over the scattered photon energy (ε_s) within the limits restricted by the Heaviside function (H) in equation (2) to get the cooling term for inverse Compton scattering (in both Thomson and Klein-Nishina regimes) of relativistic electrons in an angle-dependent photon field as

$$\frac{\mathrm{d}\gamma}{\mathrm{dt}}(\mu, \gamma, \theta) = A \left[\varepsilon_{s_{max}}^2 - \varepsilon_{s_{min}}^2 + \frac{\varepsilon_{s_{max}}^4 - \varepsilon_{s_{min}}^4}{4\gamma^2} + \frac{\varepsilon_{s_{max}}^5 - \varepsilon_{s_{min}}^5}{5\gamma^3} - \frac{2}{2.7\theta(1-\beta\mu)} \left\{ \frac{\varepsilon_{s_{max}}^3 - \varepsilon_{s_{min}}^3}{3\gamma^2} \right. \right. \tag{3}$$
$$\left. + \frac{\varepsilon_{s_{max}}^4 - \varepsilon_{s_{min}}^4}{4\gamma^3} \right\} + \frac{1}{\gamma^4} \frac{1}{(2.7\theta)^2 (1 - \beta\mu)^2} \left\{ \frac{\varepsilon_{s_{max}}^4 - \varepsilon_{s_{min}}^4}{4} + \frac{2\left(\varepsilon_{s_{max}}^5 - \varepsilon_{s_{min}}^5\right)}{5\gamma} \right.$$
$$\left. \left. + \frac{\varepsilon_{s_{max}}^6 - \varepsilon_{s_{min}}^6}{2\gamma^2} + \frac{4\left(\varepsilon_{s_{max}}^7 - \varepsilon_{s_{min}}^7\right)}{7\gamma^3} \right\} \right],$$

where $A \equiv \frac{0.3\Gamma_{rel}^2 \, c \, \sigma_T K \theta^2}{\gamma^2}$, $\varepsilon_{s_{max}} \equiv \frac{5.4\gamma^2(1-\beta\mu)\theta}{1+5.4\gamma\theta(1-\beta\mu)}$, $\varepsilon_{s_{min}} \equiv \frac{2.7\theta(1-\beta\mu)}{2}$ and $\mu \equiv \cos\psi$. ψ is the angle between the directions of propagation of the interacting photon and electron.

Figure 2 demonstrates the effect of IC cooling on electron spectra due to UV photons (i.e., $\theta = 10^{-5}$). The electron spectra evolve from an initial Maxwellian to non-Maxwellian plus a remnant of the initial Maxwellian spectra. At steady state, the spectra peak at around $\gamma \cong \Gamma \mathrm{m_i}/\mathrm{m_e}$. The IC cooling depends on the radiation density and the Lorentz factor of electrons. At lower electron energies, the effect of the cooling is almost non-existent.

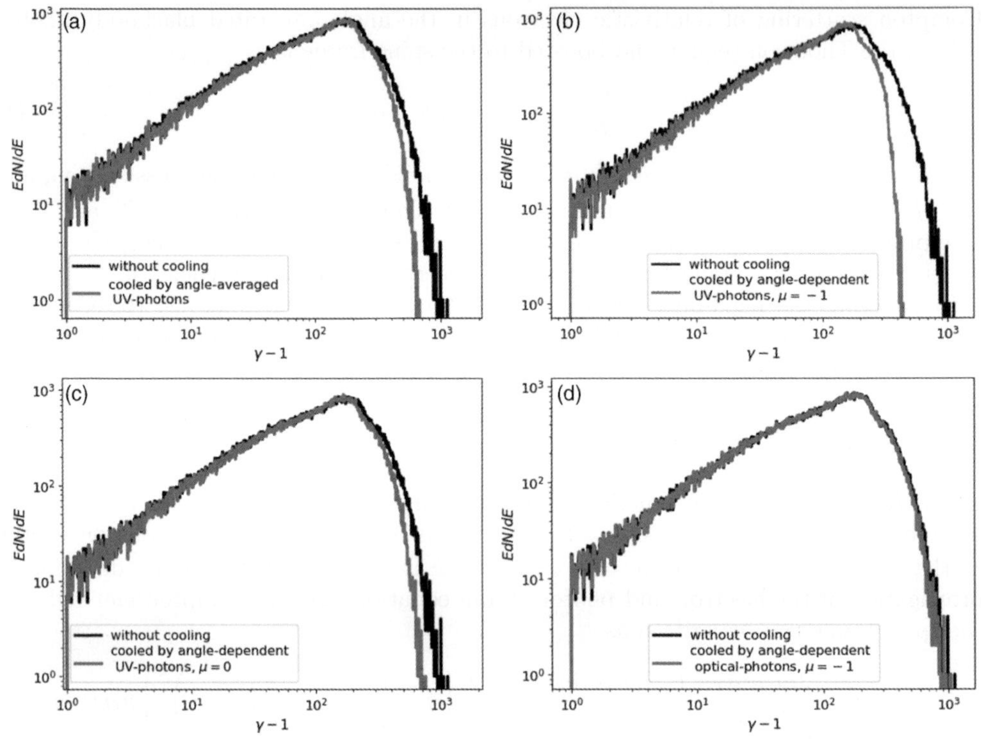

Figure 2. Comparison of Particle spectra with and without IC cooling due to UV-photons (panels a, b, and, c), and optical photons (panel d): spectra are produced at $t = 3000/\omega_{p,e}$.

However, IC cooling strongly affects electron acceleration and the electron energy spectrum at higher energies. As a result of radiation drag, the high-energy end of the spectrum shifts towards low-energies (see fig. 2). Unlike the IC cooling due to UV photons, the IC cooling strength for photons with temperature $\theta \leq 10^{-6}$ is highly subdominant. As a result, the particle dynamics and the electron energy spectrum are not appreciably affected by Compton drag.

4. Conclusion

In this paper, we discussed the results of two-dimensional PiC simulations of pure electron-ion plasmas at SBLs in relativistic jets with no initial magnetic fields, subject to IC cooling. We considered both angle-averaged and angle-dependent distribution of target photons with a δ-function approximation employed for the target photon field. We demonstrated that the relativistic particles accelerated at SBLs in relativistic jets may be substantially affected due to IC cooling. Due to the strong IC losses, high-energy electrons get cooled down to slightly lower energies. The effect of IC cooling is significantly high for the inverse Compton scattering by relativistic electrons off blackbody photons of higher temperatures (i.e., $\theta \geq 10^{-6}$). However, the highest energy end of the spectrum is slightly shifted toward lower energies. The particle spectrum is roughly unchanged for the lower temperatures of target photons (i.e., $\theta < 10^{-6}$). In the future, it may also be useful to investigate how the results change if we adopt an electron-proton plasma with a non-zero magnetic field at SBLs in relativistic jets.

Acknowledgments

This work is supported through the South African Research Chair Initiative (SARChI) of the Department of Science and Technology and the National Research Foundation† (NRF) of South Africa under SARChI chair grant no. 64789.

References

Alves, E. P., Grismayer, T., Martins, S. F., et al. 2012, The Astrophysical Journal Letters, 746, L14, doi: 10.1088/2041-8205/746/2/L14

Boettcher, M., Harris, D.E., & Krawczynski H.: 2012, Relativistic Jets from Active Galactic Nuclei

Boyd, T. M., Boyd, T., & Sanderson, J. 2003, The physics of plasmas (Cambridge university press)

Chand, T. B., Böttcher, M., & Kilian, P. 2019, PoS, HEASA2018, 046

Ghisellini, G., Tavecchio, F., & Chiaberge, M. 2005, A&A, 432, 401, doi: 10.1051/0004-6361:2004140410.48550/arXiv.astro-ph/0406093

Liang, E., Boettcher, M., & Smith, I. 2013, The Astrophysical Journal, 766, L19

Liang, E., Fu, W., Boettcher, M., Smith, I., & Roustazadeh, P. 2013, The Astrophysical Journal Letters, 779, L27, doi: 10.1088/2041-8205/779/2/L27

Owen, F. N., Hardee, P. E., & Cornwell, T. J. 1989, ApJ, 340, 698, doi: 10.1086/167430

Spitkovsky, A. 2005, AIP Conference Proceedings, doi: 10.1063/1.2141897

Weibel, E. S. 1959, Phys. Rev. Lett., 2, 83, doi: 10.1103/PhysRevLett.2.83

† Any opinion, finding and conclusion, or recommendation expressed in this material is that of the authors, and the NRF does not accept any liability in this regard.

The Multimessenger Chakra of Blazar Jets
Proceedings IAU Symposium No. 375, 2023
I. Liodakis, M. F. Aller, H. Krawczynski, A. Lähteenmäki & T. J. Pearson, eds.
doi:10.1017/S1743921323000935

Particle acceleration via magnetic reconnection in large-scale MHD jet simulations

Matteo Nurisso[1], Annalisa Celotti[1,2,3], Andrea Mignone[4] and Gianluigi Bodo[5]

[1]SISSA, Via Bonomea 265, I-34136 Trieste, Italy

[2]INAF - Osservatorio Astronomico di Brera, Via Bianchi 46, I-23807 Merate Italy,

[3]INFN - Via Valerio 2, I-34127 Trieste, Italy

[4]Dipartimento di Fisica Generale, Universitá di Torino, Via Pietro Giuria 1, I-10125 Torino, Italy

[5]INAF - Osservatorio Astrofisico di Torino, Strada Osservatorio 20, I-10025 Pino Torinese, Italy

Abstract. We present a new algorithm for the identification and physical characterization of current sheets and reconnection sites as well as the update of post-reconnection particles spectra in 2D and 3D large scale relativistic magnetohydrodynamic simulations. Lagrangian particles, which follow the fluid, are used to sample plasma parameters before entering the reconnection sites that form during the evolution of the different configurations considered. With the sampled parameters and a subgrid model based on results of Particle-in-Cell simulations we introduced in the PLUTO code an algorithm able to describe the post-reconnection spectra associated to the non-thermal component.

Keywords. magnetic reconnection, radiation mechanism: non-thermal, MHD

Magnetic reconnection (MR) is a plasma process that dissipates the energy stored in magnetic field, through a rearrangement of magnetic field topology, resulting in particles heating and acceleration. It is thought to play an important role in astrophysical sources like jets from active galactic nuclei (AGNs). MR has been studied with Particle-in-Cell (PIC) simulations, obtaining a description of the final spectrum based on plasma quantities that particles experience entering a MR region. Cold magnetization σ and the ratio between plasma and magnetic pressure β result to be the fundamental parameters to determine the resulting post-reconnection particle distribution and the acceleration efficiency.

In Nurisso *et al.* (2023) we developed a novel method to identify current sheets and to sample σ and β in magnetohydrodynamic (MHD) simulations in the Lagrangian Particle (macroparticle) module in the PLUTO code. Outside acceleration sites macroparticles spectra evolve accordingly to the relativistic cosmic–ray transport equation while MHD equations are solved concurrently, using the fluid velocity to evolve their position. While the macroparticles move towards and enter a MR region the plasma quantities necessary to infer the post-reconnection particles spectrum are sampled.

The macroparticles spectra update post-MR is implemented in Nurisso *et al.* (2023). The amount of magnetic energy available for the electron non-thermal population is estimated and the spectra are updated according to the sampled σ and β values. Spectra are then convolved with the pre-existing ones so that multiple acceleration events can be considered consistently. The description of MR in the Lagrangian Particles module constitutes, for the first time in large scale MHD simulations, a framework to study, with spatial and time resolution, acceleration due to MR in astrophysical sources.

References

Nurisso, M., Celotti, A., Mignone, A., Bodo, G., 2023, MNRAS, 522, 5517

Nurisso, M., Celotti, A., Mignone, A., Bodo, G., 2023 *In preparation*

The Multimessenger Chakra of Blazar Jets
Proceedings IAU Symposium No. 375, 2023
I. Liodakis, M. F. Aller, H. Krawczynski, A. Lähteenmäki & T. J. Pearson, eds.
doi:10.1017/S1743921323000947

Radio Polarization: A Powerful Resource for Understanding the Blazar Divide

Janhavi Baghel[1], **P. Kharb**[1], **Silpa S.**[1], **Luis C. Ho**[2,3] **and C. M. Harrison**[4]

[1]National Centre for Radio Astrophysics (NCRA) - Tata Institute of Fundamental Research (TIFR), S. P. Pune University Campus, Ganeshkhind, Pune 411007, India

[2]Kavli Institute for Astronomy and Astrophysics, Peking University, Beijing 100871, China

[3]Department of Astronomy, School of Physics, Peking University, Beijing 100871, China

[4]School of Mathematics, Statistics and Physics, Newcastle University, Newcastle upon Tyne NE1 7RU, UK

Abstract. With high-sensitivity kiloparsec-scale radio polarimetry, we can examine the jet-medium interactions and get a better understanding of the blazar divide in radio-loud (RL) AGN. We are analyzing the radio polarimetric observations with the EVLA and GMRT of 24 quasars and BL Lacs belonging to the Palomar-Green (PG) sample. The RL quasars show extensive polarisation structures in their cores, jets, lobes, and hotspots, whereas preliminary results suggest that BL Lacs exhibit polarisation primarily in their cores and inner jet regions. These findings imply that both intrinsic (central engine-related) and extrinsic (environment-related) variables are important in the formation of the blazar subclasses. The Fanaroff-Riley (FR) dichotomy can also be studied assuming RL unification and looking through the lens of blazars. Due to the radio-unbiased nature of the optically/UV-selected PG sample, we find a large fraction of the PG quasars are restarted, distorted (S- or X-shaped), or have a hybrid FR morphology.

Keywords. galaxies: active – (galaxies:) quasars: general – (galaxies:) BL Lacertae objects: general – galaxies: jets – techniques: interferometric – techniques: polarimetric

1. Introduction

Active galactic nuclei (AGN) are supermassive black holes (SMBH; Rees 1984) actively accreting matter to form the highly energetic, bright, non-stellar centres of galaxies. Relativistic jets of synchrotron-emitting plasma are sometimes launched orthogonal to the accretion disks (Blandford and Rees 1974). About 10 – 15% of AGN are radio-loud (RL) AGN parameterised as having $R = (S_{5\ GHz}/S_{B-band}) \geq 10$ where R is the ratio of radio ($5\ GHz$) to optical (B-band) flux densities (Kellermann et al. 1989). They have large-scale jets which can extend up to megaparsec scales and significantly affect their host galaxy and surroundings, influencing star formation, heating up the circumgalactic medium, and galaxy evolution (Blandford et al. 2019). Differences in observational properties have led to an intrinsically anisotropic model of the physical structure of AGN. The orientation-based RL unification hypothesis (Urry and Padovani 1995) suggests that these differences can be explained by changes in the angle at which an observer sees the AGN, obscuration by an opaque dust ring (called torus; Antonucci and Miller 1985) and the impact of relativistic beaming. Different classes of RL AGN, viz., radio galaxies (RGs), quasars and BL Lac objects, distinguished primarily by their optical emission lines, are thought to have their jets incrementally more aligned to our line of sight.

Besides orientation, other differences also exist. Fanaroff and Riley (1974) were the first to note a morphological divide in radio jets of RGs. The lower-luminosity Fanaroff-Riley type I (FRI) RGs exhibit plume-like radio lobes whereas the higher-luminosity Fanaroff-Riley type II (FRII) RGs exhibit collimated jets with terminal "hotspots". A division in total radio power was found to occur at $L_{178\ MHz} = 2 \times 10^{25}\ W\ Hz^{-1} sr^{-1}$ by Fanaroff and Riley but has lately been disputed (e.g., Mingo et al. 2019). Apart from total jet power, the combined effects of radiative losses, particle composition, and environmental effects on kiloparsec scales, are now being used to distinguish between the two classes (Croston et al. 2018). The circumstances that lead to the formation of an FRI or an FRII-like jet are still unknown. RL unification hypothesizes that quasars are the pole-on counterparts of FRII RGs while BL Lac objects are the pole-on counterparts of FRI RGs (Urry and Padovani 1995). BL Lac objects and radio-loud quasars are collectively referred to as blazars. Because of their high flux densities, they are the best sources to check the differences between the two classes with polarization observations.

1.1. *Differences in Polarisation*

Helical magnetic (B-) fields propagating outward with the jet plasma are predicted by several theoretical models of AGN jets (Meier et al. 2001; Lyutikov et al. 2005; Hawley et al. 2015). Magnetic fields are believed to play a critical role in bulk acceleration and jet propagation and are important to understand and differentiate between different jet formation mechanisms. Early polarization studies of AGN jets (e.g., Bridle and Perley 1984) have found differences between the two FR types. They found that the inferred B-field structures (deduced to be perpendicular to the electric vector position angles, EVPA, for optically thin emission, and parallel to the EVPA for optically thick emission; Pacholczyk 1970) could be categorised as (i) B-fields predominantly parallel to the jet axis, (ii) B-fields predominantly perpendicular to the jet axis, and, (iii) B-fields perpendicular to the jet axis at the centre of the jet, but parallel near one or both of its edges.

FRII sources tend to have B-fields parallel to the jet axis whereas FRI sources typically display jets with the other two B-field configurations (Willis et al. 1981; Bridle 1982). In the jets of the most powerful quasars, the inferred B-field direction after correction for Faraday rotation is normally along the jet, often following bends in the jet very closely (Bridle et al. 1994). Very long baseline interferometry (VLBI) observations of blazars have found differences in parsec scale B-field structures as well (Cawthorne et al. 1993; Kharb et al. 2008). BL Lacs tend to have their parsec scale EVPA parallel to the jet direction whereas RL quasars show EVPA perpendicular to the jet direction (Lister and Homan 2005; Lister et al. 2013). Also, BL Lacs show systematically higher parsec-scale rotation measures (RM) than the quasars (Zavala and Taylor 2005). The distances to which such B-field structures persist are still unknown and are critical to understanding the interaction between jets and their environments at kiloparsec-scales where FRI jets decelerate (Bicknell 1994; Laing and Bridle 2002).

1.2. *Kiloparsec-scale Radio Polarization Observations*

We have been carrying out polarization-sensitive kiloparsec-scale radio observations with the upgraded Giant Metrewave Radio Telescope (uGMRT) and the Karl G. Jansky Very Large Array (VLA) for a sample of blazars to study their kiloparsec-scale B-field structures. The sample chosen for this study is a subsample of the Palomar Green UV-excess photographic survey carried out in the late 1970s (Green et al. 1986). The PG

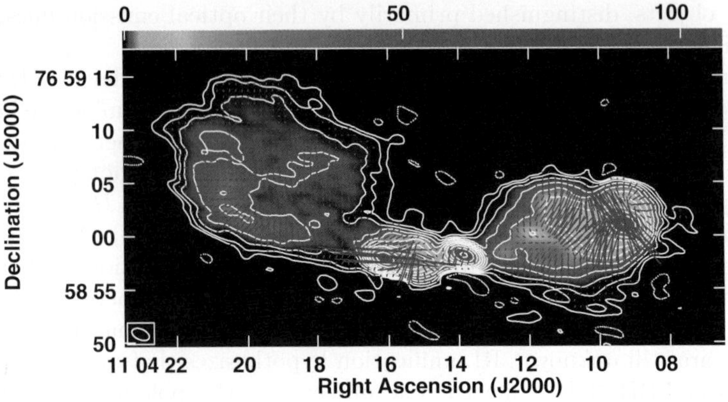

Figure 1. VLA 6 GHz total intensity contour (and color) image of quasar PG1100+772 superimposed with red EVPA vectors. The inferred B-fields are perpendicular to the EVPA vectors assuming optically thin emission. The beam is of size $1.79'' \times 1.00''$ with a PA of 67.35°. The peak surface brightness, I_P is 107.6 $mJy\ beam^{-1}$. The contour levels in percentage of I_P are (−0.09, 0.09, 0.18, 0.35, 0.7, 1.4, 2.8, 5.6, 11.25, 22.5, 45, 90) $Jy\ beam^{-1}$. The length of the EVPA vectors is proportional to polarized intensity with $5''$ corresponding to 3.57 $mJy\ beam^{-1}$ (Baghel et al. 2023).

UV-excess survey has no radio selection biases and is still the most extensive complete optically selected survey for unobscured AGN at low redshift. This makes it a useful sample to analyse the FR dichotomy, which was first noted in the radio-selected 3C sample. The PG sample is one of the most thoroughly researched AGN samples, (Boroson and Green 1992; Kellermann et al. 1989; Miller et al. 1993) and it has a wealth of supplemental multi-band data that include precise black hole (BH) masses (Kaspi et al. 2000; Wu et al. 2009; Shangguan and Ho 2019), accretion rates (Davis and Laor 2011), and galactic properties like star-formation rates (SFR) (Shi et al. 2014; Xie et al. 2021) and host galaxy morphologies (Kim et al. 2017).

The PG "blazar" sample (see Baghel et al. 2023) comprises 16 RL quasars and 8 BL Lac objects chosen using the following selection criteria:

(1) Redshift $z < 0.5$
(2) Projected core-to-lobe extents $\gtrsim 15''$

The projected lobe extents were selected to allow at least three samplings of the lobes by $\sim 5''$ uGMRT Band 4 ($\sim 650\ MHz$) observations. The $\sim 1''$ VLA 6 GHz B-array observations as a result probe regions of at least $\sim 5\ kpc$ scale at these distances. Below, we present the results from the VLA imaging of nine of the quasars and preliminary results from the uGMRT images of the BL Lac objects.

2. Results

We find that several PG quasars display distorted (S- or X-shaped) morphologies, hybrid FR radio morphologies, or restarted AGN activity (Baghel et al. 2023). We attribute this to the PG sample's optical/UV selection criteria, which remain unbiased at radio frequencies. VLA observations find evidence of ordered B-fields on kiloparsec-scales in the PG quasars. The linear fractional polarization ranges from 1-10% in the radio cores to 10-40% in the jets or lobes. The radio cores typically exhibit B-fields transverse to, and jets display B-fields typically aligned with, the jet directions. Hotspots exhibit either transverse B-field signifying compression at terminal shocks or complex B-field structures. BL Lacs and quasars do not fit neatly into the FR total power regimes at 150 MHz for the PG "blazar" sample. The VLA in-band (4.5 - 6.6 GHz) spectral indices of the cores are relatively flat but are steep in hotspots, consistent with polarization structures where the hotspots appear to be locations of jet bends or bow-shocks. Preliminary results from the uGMRT measurements at $\sim 650\ MHz$ reveal more diffuse polarised emission in blazars and indicate differences in the kiloparsec-scale B-field structures of quasars and BL Lacs.

3. Discussion and Conclusions

The distorted/hybrid/restarted radio structures could be a consequence of the selection criteria of the PG sample which relies on UV-excess and a star-like nucleus. These features are strongly tied to the accretion disk state and are basically unbiased in terms of the radio properties of these sources. High sensitivity, low-frequency radio data with less restrictive selection effects than previous studies also suggest a more complicated extended source population, which includes possible hybrid radio galaxies, restarting and remnant radio galaxies (e.g., Kapińska et al. 2017; Mingo et al. 2019; Jurlin et al. 2021). In principle, this high level of diversity in radio morphology could imply that kiloparsec-scale radio emission is highly influenced by interaction with its environment. We find that the entire PG "blazar" sample, including all the subtypes, does not cleanly fit into the FR total power regimes at 150 MHz, as well as FR radio morphologies. Several quasar lobes lack the terminal hotspot that is characteristic of FRII radio galaxies (e.g., Kharb et al. 2010).

A significant portion of our sources are radio-intermediate (with $10 < R < 250$; Falcke et al. 1996), however we find that extensive polarization is seen in these sources up to kiloparsec scales. Assuming that the aligned B-fields in the jets signify the dominance of a poloidal B-field component in them, and the transverse B-fields in the cores signify a dominant toroidal B-field component at the unresolved bases of the jets, then we see a transition from a toroidal field to a poloidal field along the radio jets (see also III Zw 2; Silpa et al. 2021). Since the poloidal field varies as r^{-2} and the toroidal field varies as $r^{-1}\ v^{-1}$, it has been hypothesized (e.g., Begelman et al. 1984) that the toroidal B-field component becomes more dominant and the poloidal component decays quicker with distance from the core. In order to explain the quasar results, one must therefore invoke a mechanism that can maintain poloidal B-fields across considerable distances, such as the cosmic battery mechanism (Contopoulos et al. 2009).

In conclusion, as higher sensitivity radio data are unveiling more complex structures in both the radio morphologies and B-fields, study of the jet-medium interactions in detail is imperative to gain a better understanding of the Fanaroff-Riley and blazar divide.

References

Antonucci, R.R.J. & Miller, J.S.: 1985, *ApJ* **297**, 621

Baghel, J. et al: 2023, *MNRAS* **519(2)**, 2773

Begelman, M.C. et al: 1984, *Rev. Mod. Phys.* **56(2)**, 255

Bicknell, G.V.: 1994, *ApJ* **422**, 542

Blandford, R. et al: 2019, *ARA&A* **57(1)**, 467

Blandford, R.D. & Rees, M.J.: 1974, *MNRAS* **169**, 395

Boroson, T.A. & Green, R.F.: 1992, *ApJS* **80**, 109

Bridle, A.H.: 1982, in D. S. Heeschen & C. M. Wade (eds.), *Extragalactic Radio Sources*, Vol. 97, pp 121–128

Bridle, A.H. et al: 1994, *AJ* **108**, 766

Bridle, A.H. & Perley, R.A.: 1984, *ARA&A* **22**, 319

Cawthorne, T.V. et al: 1993, *ApJ* **416**, 519
Contopoulos, I. et al: 2009, *ApJL* **702(2)**, L148
Croston, J.H. et al: 2018, *MNRAS* **476(2)**, 1614
Davis, S.W. & Laor, A.: 2011, *ApJ* **728(2)**, 98
Falcke, H. et al: 1996, *ApJ* **471**, 106
Fanaroff, B.L. & Riley, J.M.: 1974, *MNRAS* **167**, 31P
Green, R.F. et al: 1986, *ApJS* **61**, 305
Hawley, J.F. et al: 2015, *Space Sci. Rev.* **191(1-4)**, 441
Jurlin, N. et al: 2021, *A&A* **653**, A110
Kapińska, A.D. et al: 2017, *AJ* **154(6)**, 253
Kaspi, S. et al: 2000, *ApJ* **533(2)**, 631
Kellermann, K.I. et al: 1989, *AJ* **98**, 1195
Kharb, P. et al: 2008, *MNRAS* **384(1)**, 230
Kharb, P. et al: 2010, *ApJ* **710(1)**, 764
Kim, M. et al: 2017, *ApJS* **232(2)**, 21
Laing, R.A. & Bridle, A.H.: 2002, *MNRAS* **336(4)**, 1161
Lister, M.L. et al: 2013, *AJ* **146(5)**, 120
Lister, M.L. & Homan, D.C.: 2005, *AJ* **130(4)**, 1389
Lyutikov, M. et al: 2005, *MNRAS* **360(3)**, 869
Meier, D.L. et al: 2001, *Science* **291(5501)**, 84
Miller, P. et al: 1993, *MNRAS* **263**, 425
Mingo, B. et al: 2019, *MNRAS* **488(2)**, 2701
Pacholczyk, A.G.: 1970, *Radio astrophysics*,
Rees, M.J.: 1984, *ARA&A* **22**, 471
Shangguan, J. & Ho, L.C.: 2019, *ApJ* **873(1)**, 90
Shi, Y. et al: 2014, *ApJS* **214(2)**, 23
Silpa, S. et al: 2021, *MNRAS* **507(1)**, 991
Urry, C.M. & Padovani, P.: 1995, *PASP* **107**, 803
Willis, A.G. et al: 1981, *A&A* **95**, 250
Wu, Z.Z. et al: 2009, *Res. A&A* **9(2)**, 168
Xie, Y. et al: 2021, *ApJ* **910(2)**, 124
Zavala, R.T. & Taylor, G.B.: 2005, *ApJL* **626(2)**, L73

The Multimessenger Chakra of Blazar Jets
Proceedings IAU Symposium No. 375, 2023
I. Liodakis, M. F. Aller, H. Krawczynski, A. Lähteenmäki & T. J. Pearson, eds.
doi:10.1017/S1743921323000649

The Microvariability and Wavelength Dependence of Polarization Vector of BL Lacertae in the Outburst 2020 to 2021

Ryo Imazawa[1], Mahito Sasada[2,3], Natsuko Hazama[1], Yasushi Fukazawa[1,4], Tatsuya Nakaoka[3], Hiroshi Akitaya[5], Koji S. Kawabata[3], Thomas Bohn[3] and Anjasha Gangopadhyay[3]

[1]Department of Physics, Graduate School of Advanced Science and Engineering, Hiroshima University, Kagamiyama, 1-3-1 Higashi-Hiroshima, Hiroshima 739-8526, Japan
email: imazawa@astro.hiroshima-u.ac.jp

[2]Department of Physics, Tokyo Institute of Technology, 2-12-1 Ookayama, Meguro-ku, Tokyo 152-8551, Japan

[3]Hiroshima Astrophysical Science Center, Hiroshima University, 1-3-1 Kagamiyama, Higashi-Hiroshima, Hiroshima 739-8526, Japan

[4]Department of Physical Science, Hiroshima University 1-3-1 Kagamiyama, Higashi-Hiroshima Hiroshima 739-8526, Japan

[5]Planetary Exploration Research Center, Chiba Institute of Technology, 2-17-1 Tsudanuma, Narashino, Chiba 275-0016, Japan

Abstract. We performed photo-polarimetry observations for the blazar BL Lacertae (BL Lac) in 2020 to 2021, in which BL Lac showed historical outburst. As a result, we obtained microvariability with a timescale about five minutes and wavelength dependence of polarization degree and angle. These results indicate multiple emission regions and turbulent magnetic field structure.

Keywords. BL Lacertae, polarimetry

1. Introduction

Blazars are a type of active galactic nucleus (AGN) whose jet direction is parallel to our line of sight (Urry & Padovani 1995). The emission region and magnetic structure of the blazar jet are still under debate, and optical polarimetry is an important method for inferring the magnetic field structure of the jet. Some blazars show microvariability, which is the variation of the flux within a day. This phenomenon indicates that the emission region is compact and free from contamination by long-term variability. However, there are few observations of their polarimetry, which is useful for discussing the magnetic field in compact emission regions. BL Lacertae (BL Lac) is known as a low frequency synchrotron peaked (LSP) blazar (Abdo et al. 2010), and it has shown the microvariability several times (Miller et al. 1989; Matsumoto et al. 1999; Abeysekara et al. 2018). It showed the historic outburst since August 2020 in GeV and optical bands. We performed polarimetric observations for the flaring period of BL Lac, which is an ideal source for polarimetric studies of microvariability because it has a high polarization ($\geq 5\%$) and with a historically bright flux.

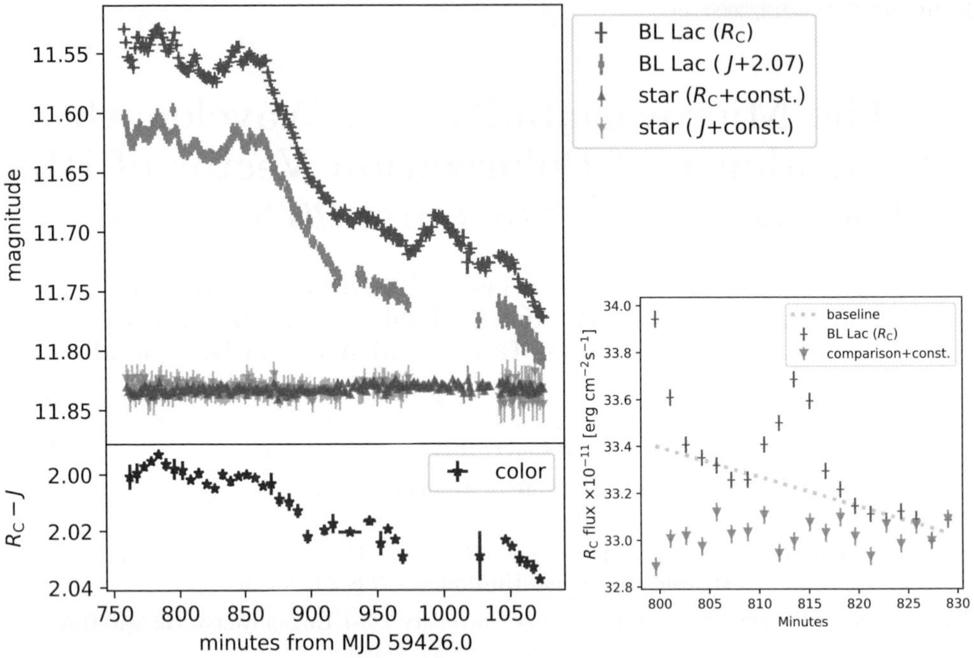

Figure 1. *Left panel*: time evolution of magnitude and color on July 31, 2021. Red crosses and gray squares in the top panel represent R_C- and J-band magnitudes, respectively. Red triangles and gray inverted-triangles indicate the comparison star. A part of the light curves in the J-band was removed due to saturation. Green stars in the bottom panel represent the R_C-J, calculated from the four-frame binned light curves. *Right panel*: light curve in the microvariability in the left panel. Red crosses and gray triangles represent the R_C band flux of BL Lac and the comparison star, respectively. The blue dashed line represents the baseline estimated by linear fitting.

2. Observations

We have performed polarimetric observations using the Hiroshima Optical and Near-InfraRed Camera (HONIR, Akitaya et al. 2014) installed on the 1.5 m Kanata telescope at the Higashi-Hiroshima Observatory, using R_C and J band filters for simultaneous observation. A fixed Wollaston prism and a rotating half-wave plate were used for polarimetry. The duration of the observations ranged from one to five hours per night and were performed over 14 nights during the bright period of BL Lac, from September 2020 to August 2021. The detailed observation and analysis method is based on Imazawa et al. (2022).

3. Results

We obtained light curves from 14 observations by considering the statistical error of 0.003–0.01 mag for the photometric data. We note that most of the days showed monotonic rise or decay with amplitudes of ~0.1 mag in a few hours. The all light curves and polarization degree and angle variations are shown in Imazawa et al. (2022). In 2021 July 31, a microvariability of the flux with a timescale of a few minutes with an amplitude 0.01–0.02 mag was found. Figure 1 shows the time variation of magnitude and R_C - J color obtained on July 31, 2021. As shown in the bottom panel of Fig 1, the color variability follows the bluer-when-brighter (BWB) trend (Clements & Carini 2001). In addition to other days in this work, the BWB trend was also seen in previous studies, which utilized long term (Wierzcholska et al. 2015) and the short term (Gaur et al. 2015) observations. The differences of the PD and PA between R_C and J band values was

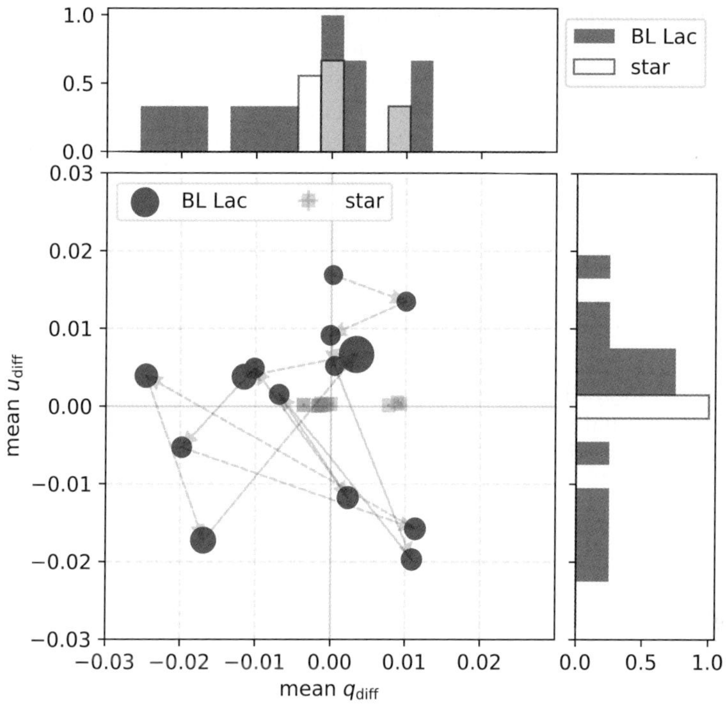

Figure 2. Scatter plot of differential q and u values between R_C and J ($q_{\rm diff}$ and $u_{\rm diff}$) for each observation. Each data point shows the average value in each observation. Gray squares represent data of the comparison star and filled circles represent data of BL Lac. The circle size represents the flux value for each data point. Gray dashed arrows that connected each data point indicate time order. Normalized histograms on the horizontal and vertical axes represent projections along the $q_{\rm diff}$ and $u_{\rm diff}$ axis, respectively.

found. Nine nights showed higher PD in the optical band than the NIR band, while five nights showed higher PD in the NIR band than in the optical band.

4. Discussion

We found variations on timescale of the order of minutes. The timescale Δt can be estimated from the following equation:

$$\Delta t = \frac{1}{1+z}\frac{\Delta F}{dF/dt}, \tag{1}$$

where z is the redshift ($z = 0.069$, Miller & Hawley 1977) and ΔF is the peak of the flux from which the baseline component is subtracted ($\Delta F = 4.60 \times 10^{-12}$ erg cm^{-2} s^{-1}). As a result of timescale estimation using linear function fitting (details in Imazawa et al. (2022)), the rise and decay times are determined to be 340 $\pm$ 45 seconds and the decay time is 278 $\pm$ 37 seconds for 31 July 2021. We confirmed that the variability is significant because the variability parameter $C/\Gamma = 5.21$ after subtracting the baseline for this variability (Romero et al. 2002; Cellone et al. 2007). We assumed that the variability timescale is 5 minutes and the maximum size of the emission region $R_{\rm em}$ is estimated to be $R_{\rm em} = (0.6 \pm 0.2) \times 10^{12}$ m.

We then discuss the Turbulent Extreme Multi-Zone (TEMZ) model, which explains the microvariability and wavelength dependence of the polarization (Marscher 2014; Marscher et al. 2017; Marscher & Jorstad 2021) by the separated multiple emission regions or ‘cells’ in the jet. In each cell, the energy spectrum of the electrons and the direction of the magnetic field are uniform. However, the energy spectrum of the electrons is different from cell to cell, and therefore the spectrum of emitted photons is also different from cell by cell. The polarized emissions from the cells cancel out with each other out due to the turbulent magnetic fields and the complicated nature of the Q and U components. Therefore, the PD from the cells is reduced depending on the number of cells. The spectral index α indicates a steep spectrum ($\alpha > 1$) and thus the number of high energy electrons should be small. Therefore, we can assume that most cells emit NIR photons while only a handful of cells emit optical photons. In this case, the PD in the NIR band should approach zero due to the large number of NIR-emitting cells, while the PD in the optical band should not be cancelled out as much due to the small number of optical-emitting cells. Remarkably, the difference in PA between the two bands can also be explained by this model (Marscher & Jorstad 2021). Therefore, the TEMZ model can better explain the high PD in the optical than in the NIR, but the reverse trend requires more complicated situations, such as the existence of accidentaty ordered magnetic field and bright cells in the NIR.

5. Summary

In this study we have performed simultaneous optical and NIR polarimetric observations for the for BL Lac during an outburst period between 2020 and 2021. The microvariability indicates the presence of compact emission regions and cannot be explained by the one-zone shock-in-jet model, but can be explained by the TEMZ model. We observed the wavelength differences of PD and PA. On several days the optical PD was higher than the NIR and vice versa. This behaviour cannot be explained by simple models, suggesting a more complicated situation such as a two-zone shock-in-jet model or an irregular TEMZ model.

Acknowledgements

This work was supported by Japan Society for the Promotion of Science (JSPS) KAKENHI Grant Numbers JP19K14761, and JP21H01137. This work was also partially supported by Optical and Near-Infrared Astronomy Inter-University Cooperation Program from the Ministry of Education, Culture, Sports, Science and Technology (MEXT) of Japan. We are grateful to the observation and operating members of Kanata Telescope. R.I. is supported by JST, the establishment of university fellowships towards the creation of science technology innovation, Grant Number JPMJFS2129. This research was supported by a grant from the Hayakawa Satio Fund awarded by the Astronomical Society of Japan.

References

Abdo, A. A., Ackermann, M., Agudo, I., et al. 2010, ApJ, 716, 30, doi: `http://doi.org/10.1088/0004-637X/716/1/30`

Abeysekara, A. U., Benbow, W., Bird, R., et al. 2018, ApJ, 856, 95, doi: `http://doi.org/10.3847/1538-4357/aab35c`

Akitaya, H., Moritani, Y., Ui, T., et al. 2014, in Society of Photo-Optical Instrumentation Engineers (SPIE) Conference Series, Vol. 9147, Ground-based and Airborne Instrumentation for Astronomy V, ed. S. K. Ramsay, I. S. McLean, & H. Takami, 91474O, doi: `http://doi.org/10.1117/12.2054577`

Cellone, S. A., Romero, G. E., & Araudo, A. T. 2007, MNRAS, 374, 357, doi: `http://doi.org/10.1111/j.1365-2966.2006.11140.x`

Clements, S. D., & Carini, M. T. 2001, AJ, 121, 90, doi: `http://doi.org/10.1086/318037`

Gaur, H., Gupta, A. C., Bachev, R., et al. 2015, MNRAS, 452, 4263, doi: `http://doi.org/10.1093/mnras/stv1556`

Imazawa, R., Sasada, M., Hazama, N., et al. 2022, pASJ, doi: http://doi.org/10.1093/pasj/psac084

Marscher, A., Jorstad, S., & Williamson, K. 2017, Galaxies, 5, 63, doi: http://doi.org/10.3390/galaxies5040063

Marscher, A. P. 2014, ApJ, 780, 87, doi: http://doi.org/10.1088/0004-637X/780/1/87

Marscher, A. P., & Jorstad, S. G. 2021, Galaxies, 9, 27, doi: http://doi.org/10.3390/galaxies9020027

Matsumoto, K., Kato, T., Nogami, D., et al. 1999, pASJ, 51, 253, doi: http://doi.org/10.1093/pasj/51.2.253

Miller, H. R., Carini, M. T., & Goodrich, B. D. 1989, Nature, 337, 627, doi: http://doi.org/10.1038/337627a0

Miller, J. S., & Hawley, S. A. 1977, ApJL, 212, L47, doi: http://doi.org/10.1086/182372

Romero, G. E., Cellone, S. A., Combi, J. A., & Andruchow, I. 2002, A&A, 390, 431, doi: http://doi.org/10.1051/0004-6361:20020743

Urry, C. M., & Padovani, P. 1995, PASP, 107, 803, doi: http://doi.org/10.1086/133630

Wierzcholska, A., Ostrowski, M., Stawarz, Ł., Wagner, S., & Hauser, M. 2015, A&A, 573, A69, doi: http://doi.org/10.1051/0004-6361/201423967

The Multimessenger Chakra of Blazar Jets
Proceedings IAU Symposium No. 375, 2023
I. Liodakis, M. F. Aller, H. Krawczynski, A. Lähteenmäki & T. J. Pearson, eds.
doi:10.1017/S1743921323000741

Intraday variations of polarization vector in blazars: a key to the optical jet structure?

Elena Shablovinskaya, Eugene Malygin and Dmitry Oparin

Special Astrophysical Observatory of RAS (SAO RAS), 369167 Nizhny Arkhyz, Russia

Abstract. This report presents the results of optical polarimetric observations carried out with 6-m and 1-m telescopes at SAO RAS. The study of the blazar S5 0716+714 radiation showed the presence of a period of the variability of brightness and polarization vector variations on scales of ~1.5 hours, constant on a long time scale; multi-colour monitoring of BL Lac polarization before, during and after the flare demonstrates the difference in the patterns of polarization vector variability depending on the wavelength. Several geometrical models and physical descriptions are discussed.

Keywords. BL Lacertae objects: BL Lac & S5 0716+714 – polarization – methods: observational

1. Introduction

For several decades, the optical variability of blazars as a special type of active galactic nuclei (AGNs) with the jet oriented towards the observer has been actively investigated. The most popular is the accumulation of long-term photometric data series (e.g. Blinov et al. 2021). However, the variability of blazars is violent and stochastic; and just as it is impossible to unambiguously reconstruct the three-dimensional air masses flow in the Earth's atmosphere according to the graph of wind speed changes over time, it is difficult to understand exclusively by the blazar light curves what physical and dynamical processes the plasma is driven by.

Fortunately, blazars are highly polarized in all spectral ranges. Their polarization of a synchrotron nature is related both to the dynamics of the emitting plasma in the unresolved jet region at scales <0.01 pc from the nucleus, and to the magnetic field in which the plasma moves. This initiates curiosity to study and map the rotation of the polarization vector, which is expected to be fast due to relativistic velocities and the small linear size of the jet.

Intraday variations (IDV) of the polarization vector direction are observed in the optical and radio bands. In the case of radio observations, IDV clearly showed the helical structure of the jet (e.g. Li et al. 2018). Patterns of optical polarization are not so clear to unambiguously reveal the same; however, long-term intraday polarization monitoring in perspective is a good tool for checking the inner models of the optical jet. Moreover, IDV continues to be little studied to date, which leaves a number of questions, e.g., how IDV depends on activity states and what physical processes dominate. Raising such issues, for several years we have been conducting intraday monitoring of the polarization of a sample of blazars. To date, we have obtained the most interesting results for two of them: S5 0716+714 and BL Lac. We briefly describe these results below.

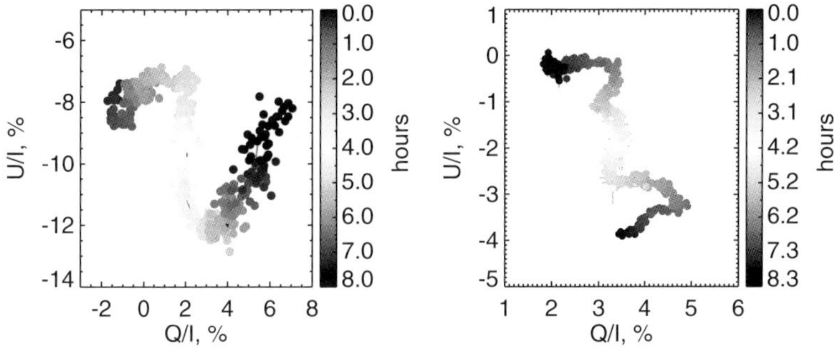

Figure 1. Overnight variations of the normalized Stokes parameters Q and U projected onto the QU-plane in S5 0716+714: 2018 (left, Shablovinskaya and Afanasiev 2019) and 2020 (right, Afanasiev et al. 2021) observation runs.

2. Brief comments on observational technique

The observations were conducted at the 6-m BTA with SCORPIO-2 (Afanasiev and Moiseev 2011) and 1-m Zeiss-1000 of SAO RAS with StoP (Afanasiev et al. 2021) and MAGIC (Afanasiev et al. 2022) devices. Regardless of the device, for polarimetric observations, we use double Wollaston prism (Oliva 1997; Geyer et al. 1993) to conduct observations in the so-called one-shot polarimetry mode unaltered by rapid atmosphere flickering. Thus, the Stokes parameters I, Q, U are measured independently for each frame. Combining it with the differential measurements by local standard stars in the field we achieve high accuracy of polarimetry (typically 0.1% for AGNs). The detailed description of prisms parameters and observational and reduction techniques could be found in (Afanasiev and Amirkhanyan 2012; Shablovinskaya and Afanasiev 2019; Shablovinskaya et al. 2023).

3. S5 0716+714 and geometrical model

S5 0716+714 is one of the brightest and most variable blazars of the northern sky, attracting special attention due to the lack of unambiguous estimates of its redshift. For the first time, we conducted its polarimetric monitoring with a 1 min cadence during the whole night on the 6-m telescope, which allowed us to detect the polarization vector changes on small time scales: switches occurred with a period of ~1.5 hours, and the same period was demonstrated by total flux variations (more details in Shablovinskaya and Afanasiev 2019). This corresponds to the cross-section size of an optical jet of the order of 10 a.u. Two years later, we repeated similar observations of S5 0716+714 on the 1-m telescope (initially for testing a new device). The analysis of the latest data confirmed the result obtained earlier at the BTA and showed that the size of the jet emitting in the optical range is stable† (Afanasiev et al. 2021). Variations of the polarization vector on the QU-plane are given in Fig. 1.

To explain the rotation of the polarization vector based on the works by Steffen et al. (1995); Nalewajko (2009), we have constructed a simple geometric model of plasma motion in a helical magnetic field in a conical jet. It is shown that the behaviour of the polarization vector on the QU-plane can be described by this model taking into account the precession of the magnetic field with a period of ~15 days. Moreover, following the calculations of Butuzova (2021), observed magnitude variations are predicted due to the Doppler factor changes. However, such a simple model is not so satisfactory because it resembles Ptolemy's epicycles: the more parameters are laid down, the more

† or, at least, typical, if we were able to detect it in observations spaced in time by several years.

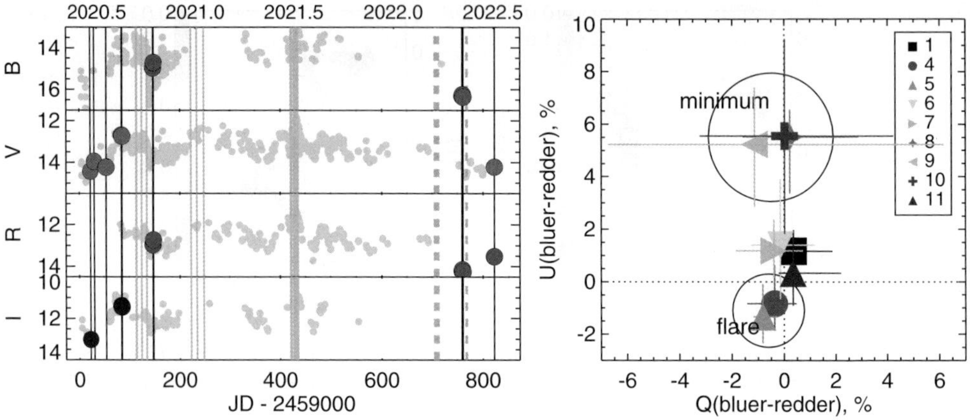

Figure 2. Left: BL Lac light curves according to AAVSO data. Our monitoring data is marked with circles. The green vertical bars indicate optical observations from (Imazawa et al. 2023), and orange dashed lines are for *IXPE* epochs from (Middei et al. 2023). Right: *QU*-diagram for the polarization difference between the "redder" and "bluer" bands (details in Shablovinskaya et al. 2023)

complex (and in general – any) pattern of polarization variations can be obtained. This prompted us to expand our observations to be able to test physical processes in plasma.

4. BL Lac and physical insights

For S5 0716+714, the *QU*-plane revealed the complex and smooth polarization IDV trajectories. But what would the picture look like in different colours? Generally speaking, due to synchrotron losses, the patterns of polarization variability should be the same, but slightly shifted in time, and this shift is proportional to the magnetic field strength. In the case of the contribution of external polarization mechanisms, the trajectories will be stably shifted relative to each other; if the physical processes in the plasma are radically different in different optical bands, then the behaviour of the polarization vector in different colours will not correlate.

To test this, we carried out observations in two optical filters (conditionally "red" and conditionally "blue") of the blazar BL Lac in 2020–2022, when the object showed extraordinary variability. Due to the activity of the blazar, other authors also observed its multi-wavelength polarization during the same period, but our epochs did not coincide, although were often close (see Fig. 2, left).

In 4 of the 9 epochs, when the blazar was in a relatively quiet state, we observed variations of brightness and polarization vector rotations without a pronounced period. The polarization chromatism was moderate, with the dominance of the "blue" component. However, during the two extreme states of the blazar, the picture was different: during the flare, the polarization vector had sharper rotation but was weaker of $\sim$1-8% and dominantly "red". During the deep minimum of the blazar, on the contrary, the variability in integral and polarized light was weak, but the maximum "blue" chromatism with a degree of polarization up to $\sim$30% was observed [see Fig. 2, on the right and (Shablovinskaya et al. 2023) for details].

The obtained observational data allow us to draw some conclusions (i.e. exclude some models) about the physical processes inside the optical jet. First, due to the polarization IDV and its chromatism on scales of an hour, assumptions about the external nature of the dependence of polarization on wavelength are excluded. In particular, any influence of the accretion disk is excluded due to the high polarization degree in the minimum state.

To explain polarization chromatism, it is possible to assume processes in a synchrotron emitting plasma, but then the electron energy distribution must be broken and change at short times. Qualitatively, the polarization behaviour can be described by the model of turbulent plasma with a shock (Marscher and Jorstad 2021) or magnetic reconnection (Zhang et al. 2022). But in order to determine which acceleration process takes place, new numerical models of both processes are required.

5. Further perspectives

By S5 0716+714 and BL Lac observations, it can be seen the polarization IDV can not only resolve the plasma motion in the optical jet but also in the future be a critical test for physical models of plasma acceleration and emission. However, this still requires more extensive statistics, including different phases of activity of blazars of different types, at larger time scales and in a wider spectral range. On the other hand, the observations demonstrate the need to combine models of synchrotron radiation, the evolution of turbulent cells and the magnetic field, etc.

Observations with the SAO RAS telescopes are supported by the Ministry of Science and Higher Education of the Russian Federation. The renovation of telescope equipment is currently provided within the national project "Science and Universities". We obtained part of the observed data on the unique scientific facility "Big Telescope Alt-azimuthal" of SAO RAS as well as made data processing with the financial support of grant No075-15-2022-262 (13.MNPMU.21.0003) of the Ministry of Science and Higher Education of the Russian Federation.

References

Afanasiev, V. L. & Amirkhanyan, V. R. 2012, Technique of polarimetric observations of faint objects at the 6-m BTA telescope. *Astrophysical Bulletin*, 67(4), 438–452.

Afanasiev, V. L., Amirkhanyan, V. R., Uklein, R. I., Perepelitsyn, A. E., Malygin, E. A., Shablovinskaya, E. S., & Afanasieva, I. V. 2022, Universal focal reducer for small telescopes. *Astronomische Nachrichten*, 343(1-2), e210104.

Afanasiev, V. L. & Moiseev, A. V. 2011, Scorpio on the 6 m Telescope: Current State and Perspectives for Spectroscopy of Galactic and Extragalactic Objects. *Baltic Astronomy*, 20, 363–370.

Afanasiev, V. L., Shablovinskaya, E. S., Uklein, R. I., & Malygin, E. A. 2021, Stokes-Polarimeter for 1-m Telescope. *Astrophysical Bulletin*, 76(1), 102–108.

Blinov, D., Kiehlmann, S., Pavlidou, V., Panopoulou, G. V., Skalidis, R., & Angelakis, E. e. a. 2021, RoboPol: AGN polarimetric monitoring data. *MNRAS*, 501(3), 3715–3726.

Butuzova, M. S. 2021, A geometrical interpretation for the properties of multiband optical variability of the blazar S5 0716+714. *Astroparticle Physics*, 129, 102577.

Geyer, E. H., Kiselev, N. N., Chernova, G. P., & Jockers, K. Surface Polarimetry of Comet Tanaka-Machholz 1992d Using a Novel Double Wollaston Prism. In *Asteroids, Comets, Meteors 1993* 1993,, volume 810, 116.

Imazawa, R., Sasada, M., Hazama, N., Fukazawa, Y., Kawabata, K. S., Nakaoka, T., Akitaya, H., Bohn, T., & Gangopadhyay, A. 2023, The microvariability and wavelength dependence of polarization degree/angle of BL Lacertae in the outburst 2020 to 2021. *Publications of the Astron. Soc. of Japan*, 75(1), 1–13.

Li, X., Mohan, P., An, T., Hong, X., Cheng, X., Yang, J., Zhang, Y., Zhang, Z., & Zhao, W. 2018, Imaging and Variability Studies of CTA 102 during the 2016 January γ-ray Flare. *ApJ*, 854(1), 17.

Marscher, A. P. & Jorstad, S. G. 2021, Frequency and Time Dependence of Linear Polarization in Turbulent Jets of Blazars. *Galaxies*, 9(2), 27.

Middei, R., Liodakis, I., Perri, M., Puccetti, S., Cavazzuti, E., Di Gesu, L., & Ehlert, e. a. 2023, X-Ray Polarization Observations of BL Lacertae. *ApJL*, 942(1), L10.

Nalewajko, K. 2009, Polarization of synchrotron emission from relativistic reconfinement shocks. *MNRAS*, 395(1), 524–530.

Oliva, E. 1997, Wedged double Wollaston, a device for single shot polarimetric measurements. *Astronomy and Astrophysics, Supplement*, 123, 589–592.

Shablovinskaya, E., Malygin, E., & Oparin, D. 2023, Chromatic optical polarization of BL Lac: while faint and bright. *MNRAS*, 519(3), 3798–3810.

Shablovinskaya, E. S. & Afanasiev, V. L. 2019, The intraday variations of the polarization vector direction in radio source S5 0716+714. *MNRAS*, 482(4), 4322–4328.

Steffen, W., Zensus, J. A., Krichbaum, T. P., Witzel, A., & Qian, S. J. 1995, A helical model for the compact jet in 3C345. *A&A*, 302, 335.

Zhang, H., Li, X., Giannios, D., Guo, F., Thiersen, H., Böttcher, M., Lewis, T., & Venters, T. 2022, Radiation and Polarization Signatures from Magnetic Reconnection in Relativistic Jets. II. Connection with γ-Rays. *ApJ*, 924(2), 90.

The Multimessenger Chakra of Blazar Jets
Proceedings IAU Symposium No. 375, 2023
I. Liodakis, M. F. Aller, H. Krawczynski, A. Lähteenmäki & T. J. Pearson, eds.
doi:10.1017/S1743921323000625

Identifying γ-ray emitting blazars in the PASIPHAE era: polarimetry as a unique probe

Nikolaos Mandarakas

Department of Physics, University of Crete, 70013, Heraklion, Greece

Institute of Astrophysics, Foundation for Research and Technology - Hellas, Voutes, 70013 Heraklion, Greece

Abstract. The Fermi γ-ray telescope has detected 6658 sources, with 1845 of them remaining unidentified. We show that polarimetry of γ-ray fields is a powerful asset in the hunt of active galactic nuclei (AGN) as potential optical counterparts for γ-ray sources. We have studied an unidentified *Fermi* field (3FGL J0221.2+2518) and found a previously-unknown highly-polarized extragalactic object as a potential optical counterpart within the 1-sigma error ellipse of the corresponding γ-ray source. Based on a collection of data, we find that it most probably is a composite object: a star-forming galaxy accompanied by AGN. PASIPHAE is a large polarimetric experiment which will measure the polarisation of sources away from the galactic plane. This will provide an excellent opportunity to study hundreds of unidentified γ-ray sources and unveil potential optical counterparts, using polarimetry.

Keywords. techniques: polarimetric; galaxies: active; gamma rays: galaxies; Astrophysics - High Energy Astrophysical Phenomena

1. Introduction

The *Fermi* spacecraft was launched in 2008 and has been scanning the sky to collect data on γ-ray sources. The latest data release (4FGL-DR3, Abdollahi et al. 2022) contains 6658 sources from the first 12 years of operation. While 1845 sources still remain unidentified, $\sim 86\%$ of identifications and associations consist of different types of active galactic nuclei (AGN).

Efforts for classifying γ-ray sources have been carried out using a variety of techniques, such as machine learning (Doert and Errando 2014; Chiaro et al. 2016; Salvetti et al. 2017), very long baseline interferometry (VLBI, Kovalev 2009), and multi-wavelength observations (Acero et al. 2013).

For the first time, we introduced optical polarimetry as a promising method to indentify AGN as potential optical counterparts for unidentified gamma-ray sources (UGSs), and proposed a new candidate for one of the *Fermi* fields (Mandarakas et al. 2019). Liodakis and Blinov (2019) utilised optical polarisation data together with machine learning techniques and identified the same source as a potential optical counterpart for the same *Fermi* field.

AGN constitute the vast majority ($\sim 86\%$) of the already indentified γ-ray sources in the *Fermi* catalogue, and are known to be highly polarized due to the synchrotron nature of their optical emission (Angel and Stockman 1980; Angelakis et al. 2016). However, not only AGN can appear polarized in the sky. In any given line of sight, the light passes through the interstellar medium (ISM) of our galaxy and becomes linearly polarized due

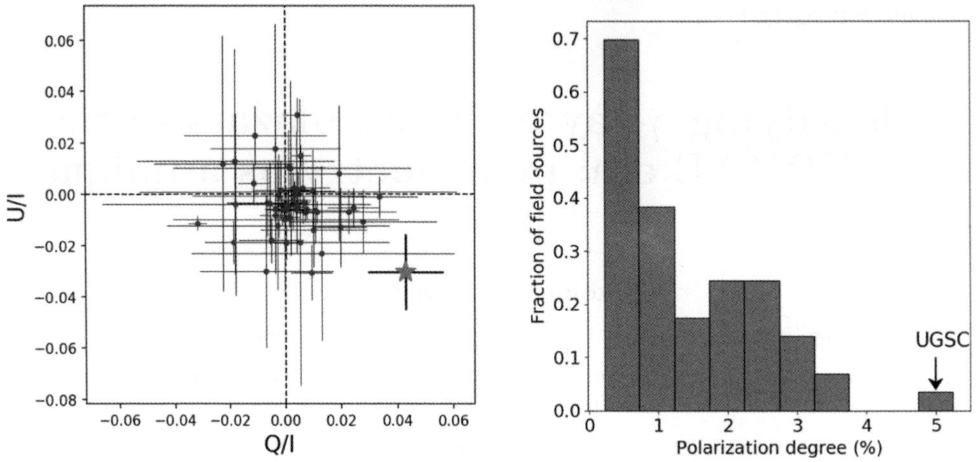

Figure 1. Left: Distribution of q and u Stokes parameters of the 3FGL J0221.2+2518 field sources. UGSC is shown by the red star. Right: Distribution of the polarisation values for the observed sources in the field.

to dichroic extinction from dust grains that are aligned with the interstellar magnetic field (Andersson et al. 2015). The ISM-induced polarisation is typically in the levels of a few percent, which can be significant enough to be compared to the intrinsic polarisation of AGN. In addition, AGN can be variable in polarisation, which can make their polarisation drop down to the ISM-induced polarisation, thus making them indistinguishable. We used data from the RoboPol survey (Blinov et al. 2021) to simulate AGN in the UGSs. We demonstrated that in high galactic latitudes, where the interstelllar polarisation is expected to be at the $\sim$ 1% level, an AGN in the field will significantly stand out in polarisation $\sim$80% of the times (for the details refer to Mandarakas et al. 2019).

2. Demonstration of the technique

In order to demonstrate our method, we used the RoboPol instrument mounted on Skinakas 1.3 m telescope†. RoboPol is a novel polarimeter with no rotating parts that can measure the linear q and u Stokes parameters with a single exposure (Ramaprakash et al. 2019). We note that at the time of this work, the latest available *Fermi* data release was the 3FGL catalogue (Acero et al. 2015). We measured the polarisation of optical sources within the 3 σ position ellipse of the 3FGL J0221.2+2518 γ-ray field. The results are demonstrated in Fig. 1, where it is evident that there is a source that significantly stands out in polarisation from the average of the field. This is our proposed Unidentified Gamma-ray Source Candidate (UGSC). The polarisation of UGSC is $5.2 \pm 1.3\%$ while the field average is $p_{av} = 0.91\%$. We obtained the spectrum of the source, using the Skinakas 1.3 m telescope and we verified that our candidate is an extragalactic source, and most probably a composite object: a starburst galaxy accompanied by an AGN (for the details refer to Mandarakas et al. 2019).

2.1. *The predictive power of polarimetry*

The UGSC is located at RA=02 21 33.3s, DEC=+25 12 47.2 (J2000), and is listed in SDSS as J022133.31+251247.3. Based on the position of the γ-ray source reported in the third data release of *Fermi* (which was the newest version available at the time of the study), UGSC lies between the 2 and 3 σ error ellipses of the γ-ray source location (left

† `https://skinakas.physics.uoc.gr/en/`

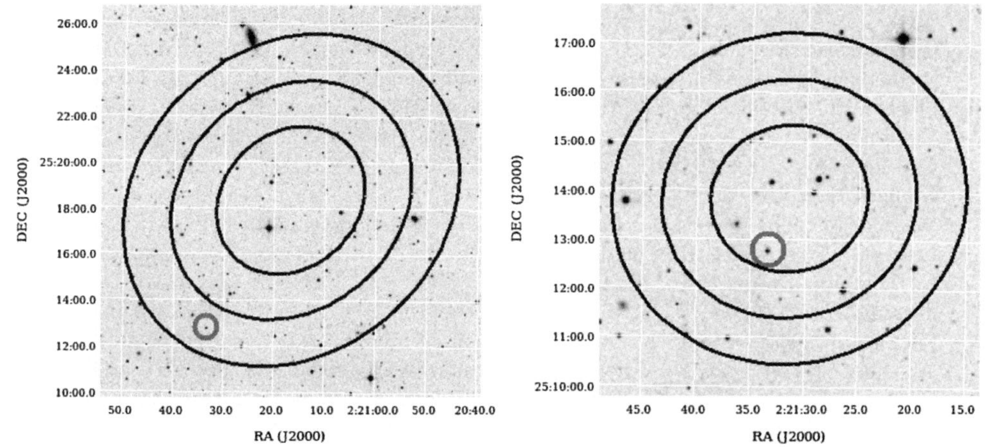

Figure 2. Left: DSS image of 3FGL J0221.2+2518 field and the position of UGSC (red circle). Right: DSS image of 4FGL J0221.5+2513 and the position of UGSC (red circle). The ellipses represent $1\,\sigma$, $2\,\sigma$, and $3\,\sigma$ uncertainties of the γ-ray source positions. The position of the γ-ray source has shifted between the two versions of the *Fermi* catalogue and has come to agree better with the polarimetric candidate. Note the scale difference. Localisation is improved in the latest version of the catalogue.

of Fig. 2). However, in the latest *Fermi* data release, where the localisation and overall analysis has been improved, our candidate lies within the $1\,\sigma$ error ellipse of the γ-ray source location (right of Fig. 2). This is indicative of the potential of polarimetry in the identification of optical counterparts of yet-unidentified *Fermi* sources: the refinement of γ-ray data led to a shift of the γ-ray source to coincide with the candidate proposed by this polarimetric study.

3. Future prospect

PASIPHAE is a groundbreaking survey that aims to scan the sky away from the galactic plane in both hemispheres and measure the polarisation of millions of objects within its range (Tassis et al. 2018). PASIPHAE will make use of two state-of-the-art wide-field polarimeters, which will split the light in four orthogonal polarisation states and project each one in a different CCD camera, leading to high-quality observations (Maharana et al. 2021, 2022). The diversity of the mapped sources will be rich, including stars, galaxies, AGN and hundreds of unidentified *Fermi* γ-ray sources. By mining the PASIPHAE data, and making use of the excellent prospect of optical polarimetry, discoveries of AGN as potential optical counterparts for γ-ray sources will come for free, and could yield hundreds of new identifications.

Acknowledgements

I acknowledge support from the European Research Council (ERC) under the European Union Horizon 2020 research and innovation program under grant agreement No 771282. The *RoboPol* project is a collaboration between Caltech in the USA, MPIfR in Germany, Toruń Centre for Astronomy in Poland, the University of Crete/FORTH in Greece, and IUCAA in India.

References

Abdollahi, S., Acero, F., Baldini, L., Ballet, J., Bastieri, D., Bellazzini, R., Berenji, B., Berretta, A., Bissaldi, E., Blandford, R. D., Bloom, E., Bonino, R., Brill, A., Britto, R. J., Bruel, P., Burnett, T. H., Buson, S., Cameron, R. A., Caputo, R., Caraveo, P. A., Castro, D.,

Chaty, S., Cheung, C. C., Chiaro, G., Cibrario, N., Ciprini, S., Coronado-Blázquez, J., Crnogorcevic, M., Cutini, S., D'Ammando, F., De Gaetano, S., Digel, S. W., Di Lalla, N., Dirirsa, F., Di Venere, L., Domínguez, A., Fallah Ramazani, V., Fegan, S. J., Ferrara, E. C., Fiori, A., Fleischhack, H., Franckowiak, A., Fukazawa, Y., Funk, S., Fusco, P., Galanti, G., Gammaldi, V., Gargano, F., Garrappa, S., Gasparrini, D., Giacchino, F., Giglietto, N., Giordano, F., Giroletti, M., Glanzman, T., Green, D., Grenier, I. A., Grondin, M. H., Guillemot, L., Guiriec, S., Gustafsson, M., Harding, A. K., Hays, E., Hewitt, J. W., Horan, D., Hou, X., Jóhannesson, G., Karwin, C., Kayanoki, T., Kerr, M., Kuss, M., Landriu, D., Larsson, S., Latronico, L., Lemoine-Goumard, M., Li, J., Liodakis, I., Longo, F., Loparco, F., Lott, B., Lubrano, P., Maldera, S., Malyshev, D., Manfreda, A., Martí-Devesa, G., Mazziotta, M. N., Mereu, I., Meyer, M., Michelson, P. F., Mirabal, N., Mitthumsiri, W., Mizuno, T., Moiseev, A. A., Monzani, M. E., Morselli, A., Moskalenko, I. V., Negro, M., Nuss, E., Omodei, N., Orienti, M., Orlando, E., Paneque, D., Pei, Z., Perkins, J. S., Persic, M., Pesce-Rollins, M., Petrosian, V., Pillera, R., Poon, H., Porter, T. A., Principe, G., Rainò, S., Rando, R., Rani, B., Razzano, M., Razzaque, S., Reimer, A., Reimer, O., Reposeur, T., Sánchez-Conde, M., Saz Parkinson, P. M., Scotton, L., Serini, D., Sgrò, C., Siskind, E. J., Smith, D. A., Spandre, G., Spinelli, P., Sueoka, K., Suson, D. J., Tajima, H., Tak, D., Thayer, J. B., Thompson, D. J., Torres, D. F., Troja, E., Valverde, J., Wood, K., & Zaharijas, G. 2022, Incremental Fermi Large Area Telescope Fourth Source Catalog. *ApJS*, 260(2), 53.

Acero, F., Ackermann, M., Ajello, M., Albert, A., Atwood, W. B., Axelsson, M., Baldini, L., Ballet, J., Barbiellini, G., Bastieri, D., Belfiore, A., Bellazzini, R., Bissaldi, E., Blandford, R. D., Bloom, E. D., Bogart, J. R., Bonino, R., Bottacini, E., Bregeon, J., Britto, R. J., Bruel, P., Buehler, R., Burnett, T. H., Buson, S., Caliandro, G. A., Cameron, R. A., Caputo, R., Caragiulo, M., Caraveo, P. A., Casandjian, J. M., Cavazzuti, E., Charles, E., Chaves, R. C. G., Chekhtman, A., Cheung, C. C., Chiang, J., Chiaro, G., Ciprini, S., Claus, R., Cohen-Tanugi, J., Cominsky, L. R., Conrad, J., Cutini, S., D'Ammando, F., de Angelis, A., DeKlotz, M., de Palma, F., Desiante, R., Digel, S. W., Di Venere, L., Drell, P. S., Dubois, R., Dumora, D., Favuzzi, C., Fegan, S. J., Ferrara, E. C., Finke, J., Franckowiak, A., Fukazawa, Y., Funk, S., Fusco, P., Gargano, F., Gasparrini, D., Giebels, B., Giglietto, N., Giommi, P., Giordano, F., Giroletti, M., Glanzman, T., Godfrey, G., Grenier, I. A., Grondin, M. H., Grove, J. E., Guillemot, L., Guiriec, S., Hadasch, D., Harding, A. K., Hays, E., Hewitt, J. W., Hill, A. B., Horan, D., Iafrate, G., Jogler, T., Jóhannesson, G., Johnson, R. P., Johnson, A. S., Johnson, T. J., Johnson, W. N., Kamae, T., Kataoka, J., Katsuta, J., Kuss, M., La Mura, G., Landriu, D., Larsson, S., Latronico, L., Lemoine-Goumard, M., Li, J., Li, L., Longo, F., Loparco, F., Lott, B., Lovellette, M. N., Lubrano, P., Madejski, G. M., Massaro, F., Mayer, M., Mazziotta, M. N., McEnery, J. E., Michelson, P. F., Mirabal, N., Mizuno, T., Moiseev, A. A., Mongelli, M., Monzani, M. E., Morselli, A., Moskalenko, I. V., Murgia, S., Nuss, E., Ohno, M., Ohsugi, T., Omodei, N., Orienti, M., Orlando, E., Ormes, J. F., Paneque, D., Panetta, J. H., Perkins, J. S., Pesce-Rollins, M., Piron, F., Pivato, G., Porter, T. A., Racusin, J. L., Rando, R., Razzano, M., Razzaque, S., Reimer, A., Reimer, O., Reposeur, T., Rochester, L. S., Romani, R. W., Salvetti, D., Sánchez-Conde, M., Saz Parkinson, P. M., Schulz, A., Siskind, E. J., Smith, D. A., Spada, F., Spandre, G., Spinelli, P., Stephens, T. E., Strong, A. W., Suson, D. J., Takahashi, H., Takahashi, T., Tanaka, Y., Thayer, J. G., Thayer, J. B., Thompson, D. J., Tibaldo, L., Tibolla, O., Torres, D. F., Torresi, E., Tosti, G., Troja, E., Van Klaveren, B., Vianello, G., Winer, B. L., Wood, K. S., Wood, M., Zimmer, S., & Fermi-LAT Collaboration 2015, Fermi Large Area Telescope Third Source Catalog. *ApJS*, 218(2), 23.

Acero, F., Donato, D., Ojha, R., Stevens, J., Edwards, P. G., Ferrara, E., Blanchard, J., Lovell, J. E. J., & Thompson, D. J. 2013, Hunting for Treasures among the Fermi Unassociated Sources: A multiwavelength Approach. *ApJ*, 779(2), 133.

Andersson, B. G., Lazarian, A., & Vaillancourt, J. E. 2015, Interstellar Dust Grain Alignment. *ARA&A*, 53, 501–539.

Angel, J. R. P. & Stockman, H. S. 1980, Optical and infrared polarization of active extragalactic objects. *ARA&A*, 18, 321–361.

Angelakis, E., Hovatta, T., Blinov, D., Pavlidou, V., Kiehlmann, S., Myserlis, I., Böttcher, M., Mao, P., Panopoulou, G. V., Liodakis, I., King, O. G., Baloković, M., Kus, A., Kylafis, N., Mahabal, A., Marecki, A., Paleologou, E., Papadakis, I., Papamastorakis, I., Pazderski, E., Pearson, T. J., Prabhudesai, S., Ramaprakash, A. N., Readhead, A. C. S., Reig, P., Tassis, K., Urry, M., & Zensus, J. A. 2016, RoboPol: the optical polarization of gamma-ray-loud and gamma-ray-quiet blazars. *MNRAS*, 463(3), 3365–3380.

Blinov, D., Kiehlmann, S., Pavlidou, V., Panopoulou, G. V., Skalidis, R., Angelakis, E., Casadio, C., Einoder, E. N., Hovatta, T., Kokolakis, K., Kougentakis, A., Kus, A., Kylafis, N., Kyritsis, E., Lalakos, A., Liodakis, I., Maharana, S., Makrydopoulou, E., Mandarakas, N., Maragkakis, G. M., Myserlis, I., Papadakis, I., Paterakis, G., Pearson, T. J., Ramaprakash, A. N., Readhead, A. C. S., Reig, P., Słowikowska, A., Tassis, K., Xexakis, K., Żejmo, M., & Zensus, J. A. 2021, RoboPol: AGN polarimetric monitoring data. *MNRAS*, 501(3), 3715–3726.

Chiaro, G., Salvetti, D., La Mura, G., Giroletti, M., Thompson, D. J., & Bastieri, D. 2016, Blazar flaring patterns (B-FlaP) classifying blazar candidate of uncertain type in the third Fermi-LAT catalogue by artificial neural networks. *MNRAS*, 462(3), 3180–3195.

Doert, M. & Errando, M. 2014, Search for Gamma-ray-emitting Active Galactic Nuclei in the Fermi-LAT Unassociated Sample Using Machine Learning. *ApJ*, 782(1), 41.

Kovalev, Y. Y. 2009, Identification of the Early Fermi/LAT Gamma-Ray Bright Objects with Extragalactic VLBI Sources. *ApJL*, 707(1), L56–L59.

Liodakis, I. & Blinov, D. 2019, Probing the unidentified Fermi blazar-like population using optical polarization and machine learning. *MNRAS*, 486(3), 3415–3422.

Maharana, S., Anche, R. M., Ramaprakash, A. N., Joshi, B., Basyrov, A., Blinov, D., Casadio, C., Deka, K., Eriksen, H. K., Ghosh, T., Gjerløw, E., Kypriotakis, J. A., Kiehlmann, S., Mandarakas, N., Panopoulou, G. V., Papadaki, K., Pavlidou, V., Pearson, T. J., Pelgrims, V., Potter, S. B., Readhead, A. C. S., Skalidis, R., Svalheim, T. L., Tassis, K., & Wehus, I. K. 2022, WALOP-South: a four-camera one-shot imaging polarimeter for PASIPHAE survey. Paper II - polarimetric modeling and calibration. *Journal of Astronomical Telescopes, Instruments, and Systems*, 8, 038004.

Maharana, S., Kypriotakis, J. A., Ramaprakash, A. N., Rajarshi, C., Anche, R. M., Shrish, Blinov, D., Eriksen, H. K., Ghosh, T., Gjerløw, E., Mandarakas, N., Panopoulou, G. V., Pavlidou, V., Pearson, T. J., Pelgrims, V., Potter, S. B., Readhead, A. C. S., Skalidis, R., Tassis, K., & Wehus, I. K. 2021, WALOP-South: a four-camera one-shot imaging polarimeter for PASIPHAE survey. Paper I—optical design. *Journal of Astronomical Telescopes, Instruments, and Systems*, 7, 014004.

Mandarakas, N., Blinov, D., Liodakis, I., Kouroumpatzakis, K., Zezas, A., Panopoulou, G. V., Myserlis, I., Angelakis, E., Hovatta, T., Kiehlmann, S., Kokolakis, K., Paleologou, E., Pouliasi, A., Skalidis, R., & Pavlidou, V. 2019, Search for AGN counterparts of unidentified Fermi-LAT sources with optical polarimetry. Demonstration of the technique. *A&A*, 623, A61.

Ramaprakash, A. N., Rajarshi, C. V., Das, H. K., Khodade, P., Modi, D., Panopoulou, G., Maharana, S., Blinov, D., Angelakis, E., Casadio, C., Fuhrmann, L., Hovatta, T., Kiehlmann, S., King, O. G., Kylafis, N., Kougentakis, A., Kus, A., Mahabal, A., Marecki, A., Myserlis, I., Paterakis, G., Paleologou, E., Liodakis, I., Papadakis, I., Papamastorakis, I., Pavlidou, V., Pazderski, E., Pearson, T. J., Readhead, A. C. S., Reig, P., Słowikowska, A., Tassis, K., & Zensus, J. A. 2019, RoboPol: a four-channel optical imaging polarimeter. *MNRAS*, 485(2), 2355–2366.

Salvetti, D., Chiaro, G., La Mura, G., & Thompson, D. J. 2017, 3FGLzoo: classifying 3FGL unassociated Fermi-LAT γ-ray sources by artificial neural networks. *MNRAS*, 470(2), 1291–1297.

Tassis, K., Ramaprakash, A. N., Readhead, A. C. S., Potter, S. B., Wehus, I. K., Panopoulou, G. V., Blinov, D., Eriksen, H. K., Hensley, B., Karakci, A., Kypriotakis, J. A., Maharana, S., Ntormousi, E., Pavlidou, V., Pearson, T. J., & Skalidis, R. 2018, PASIPHAE: A high-Galactic-latitude, high-accuracy optopolarimetric survey. *arXiv e-prints*,, arXiv:1810.05652.

The Multimessenger Chakra of Blazar Jets
Proceedings IAU Symposium No. 375, 2023
I. Liodakis, M. F. Aller, H. Krawczynski, A. Lähteenmäki & T. J. Pearson, eds.
doi:10.1017/S1743921323000881

Scientific Potential of MeV Polarimetry for Relativistic Jets

Haocheng Zhang

NASA Postdoc Program Fellow haocheng.zhang@nasa.gov
NASA Goddard Space Flight Center, Greenbelt, MD 20771, USA

Abstract. Relativistic jets from supermassive black holes or stellar mass black holes are among the most powerful astrophysical phenomena. Magnetic field plays an important role in the jet launching and propagation, as well as particle acceleration and radiation. Polarimetry is the only way to observe the magnetic field evolution. The recent launch of the Imaging X-ray Polarimetry Explorer (IXPE) has opened up the X-ray polarization window, which has revealed very interesting phenomena for relativistic jets. However, the field of MeV gamma-ray polarimetry remains largely unexplored. This paper aims to summarize key scientific potentials for MeV polarimetry for blazars and gamma-ray bursts (GRBs) from recent theoretical modeling. These predictions, which are closely related to the cosmic ray acceleration, neutrino production, radiation mechanism, and the jet evolution, can be examined by future MeV polarimeters, such as the Compton Spectrometer and Imager (COSI), the LargE Area burst Polarimeter (LEAP), and the All-sky Medium-Energy Gamma-ray Observatory eXplorer (AMEGO-X).

Keywords. polarimetry, blazar, gamma-ray burst

1. Introduction

Relativistic jets are highly energetic plasma outflows from accreting black holes. If the jet points very close to our line of sight, its apparent luminosity is boosted by the relativistic beaming effects. As a result, blazars, jets from supermassive black holes, and GRBs, jets from stellar mass black holes, are among the most powerful extragalactic γ-ray sources detected by *Fermi* (Abdollahi et al. 2020). Both blazars and GRBs are highly variable multi-wavelength sources, indicating extreme particle acceleration in the jet.

The highly variable multi-wavelength blazar emission shows a characteristic two-component spectrum. The low-energy spectral component extends from radio to optical bands, in some cases up to soft X-rays, while the high-energy component covers the X-ray and γ-ray bands. The low-energy component is dominated by the synchrotron emission from nonthermal electrons, evident by the observed high radio and optical polarization degree (for a recent review, see Zhang 2019), thus it is often referred to as the synchrotron component. Depending on the spectral peak of this component, blazars can be classified as the low-synchrotron-peaked (LSP), whose synchrotron peak is in the infrared bands, intermediate-synchrotron peaked (ISP), whose peak is in the optical to ultraviolet bands, and high-synchrotron-peaked (HSP), which peaks in the soft X-rays (Abdo et al. 2010). Observations have found that the location of the synchrotron peak is correlated to other observational properties such as the blazar luminosity and the spectral hardness (e.g., Ajello et al. 2020). This correlation is often referred to as the "blazar sequence". The origin of the high-energy component is still under debate. Many theoretical models suggest that the high-energy component results from the inverse Compton scattering of low-energy photons by the same nonthermal electron population that produces the synchrotron component (Dermer et al. 1992; Ghisellini & Madau 1996; Maraschi et al. 1992; Marscher & Gear 1985; Sikora et al. 1994). The seed photons for Compton scattering can be the synchrotron emission itself, called the synchrotron self Compton (SSC), or the external thermal photons from the accretion disk, broad line region, molecular cloud, and/or the dusty torus, which is referred to as the external Compton (EC). This leptonic scenario is supported by the frequently observed simultaneous flaring of the synchrotron and high-energy components. On the other hand, the hadronic scenario argues that the blazar jet is sufficiently powerful to accelerate very high energy protons, which can produce the high-energy component via proton synchrotron and/or hadronic cascades (Aharonian 2000; Mannheim 1993; Mücke et al. 2003). Under this scenario, blazars are the extragalactic origin of cosmic rays and neutrinos (Böttcher et al. 2013; Cerruti et al. 2015). Recently, the *IceCube* detection of a very high energy neutrino in coincidence with the blazar TXS 0506+056 flare provides the first evidence for the hadronic scenario (IceCube Collaboration et al. 2018). However, detection of extragalactic neutrinos is very hard, and the leptonic and hadronic models generally produce similar multi-wavelength spectra, requiring additional observable signatures to distinguish between the two models (Chen et al. 2014; Diltz et al. 2015; Li & Kusunose 2000; Petropoulou et al. 2015; Sahu et al. 2013, and see Böttcher (2019) for a recent review).

GRBs consists of long bursts (more than two seconds) and short bursts (less than two seconds). Their origins are generally believed to be different: the former is probably due to the collapse of massive stars, while the latter results from the merger of binary neutron stars (Abbott et al. 2017; Goldstein et al. 2017; Hjorth et al. 2003; MacFadyen et al. 2001). For either long or short bursts, the prompt phase of the GRB is followed by the afterglow, which is due to the synchrotron emission in an optically thin environment. The prompt phase spectrum can be described by the Band function, but the underlying radiation mechanism is not well understood. Both the photospheric radiation and synchrotron emission in an optically thin environment are considered. In the photospheric model, photons are emitted inside the jet and keep interacting with the jet materials until they finally escape when the optical depth $\tau \sim 1$ (Beloborodov 2010; Pe'er et al. 2006; Rees & Mészáros 2005). In the synchrotron model, theoretical models such as the internal-collision-induced magnetic reconnection and turbulence (ICMART) model and the striped jet model often consider that the prompt emission region is magnetized, and the nonthermal particles are accelerated via magnetic reconnection and/or turbulence (Giannios & Uzdensky 2019; Rees & Meszaros 1994; Zhang & Yan 2011). The magnetic field in the prompt emission region may be ordered or disordered under the synchrotron model (Toma et al. 2009; Zhang & Yan 2011). Similar to the blazar γ-ray models, both the photospheric and synchrotron models can describe the GRB prompt emission spectra and light curves reasonably, requiring additional constraints to distinguish the two models.

This paper presents a review on the scientific potentials of MeV polarimetry for blazars and GRBs. For both systems, MeV polarimetry can diagnose the radiation mechanism of the γ-ray emission. MeV polarization degree that is comparable to or higher than the optical counterpart in blazars is an unambiguous evidence for the hadronic model, which can constrain the neutrino production and maximal cosmic ray energy obtainable in the blazar emission region. Additionally, the MeV polarization degree does not vary quickly due to the slow cooling of protons, ideal for detection by future MeV polarimeters. On the other hand, while a high MeV polarization degree prefers synchrotron emission in an ordered magnetic field for GRBs, the photospheric model cannot be ruled out. Time-dependent MeV polarization degree and angle variations are necessary to fully distinguish the physical mechanism for the GRB prompt emission.

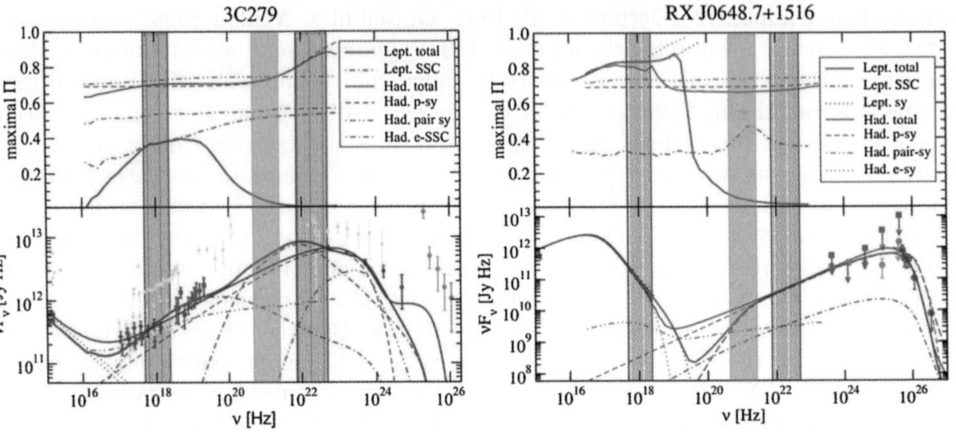

Figure 1. Multi-wavelength maximal polarization degree (upper panels) and spectral energy distribution (lower panels) for the two blazars 3C279 (left) and RX J0648.7+1516 (right). Leptonic model fits are plotted in red, hadronic models in green. Solid lines represent the model total, while other line styles indicate individual radiation components, as labeled in the legend. The orange shaded region represents the MeV band, while the blue shaded region on the left is approximately the *IXPE* soft X-ray band. This figure is reproduced from Zhang & Böttcher (2013).

2. Probe Blazar Cosmic Rays and Neutrinos with MeV Polarimetry

Neutrinos are obviously the smoking gun of the hadronic processes in blazar jets, as shown by the recent *IceCube* event (IceCube Collaboration et al. 2018). However, current neutrino detectors such as *IceCube* obtain adequately high statistics for neutrino detection. Additionally, the angular resolution is not very good, making it hard to determine the origin of the neutrinos. Given that neutrinos can travel through the universe without significant interaction with other materials, it is ambiguous to show whether neutrinos really come from a specific flaring blazar, even if the two events are simultaneous. MeV polarimetry provides a novel way to probe hadronic processes in blazars by identifying the radiation mechanisms. Aforementioned, the high-energy blazar spectral component in the hadronic model is dominated by the synchrotron emission from either protons or secondary charged particles produced by the hadronic cascades. The synchrotron emission is characterized by high polarization. In a perfectly ordered magnetic field, synchrotron emission can reach $\gtrsim 70\%$ polarization degree, depending on the spectral index (see Rybicki & Lightman 1986, for a detailed derivation). On the other hand, the inverse Compton scattering, which dominates the high-energy spectral component under the leptonic model, has relatively low polarization. Bonometto et al. (1970); Bonometto & Saggion (1973) have shown that isotropic inverse Compton scattering roughly reduces by half the polarization of the seed photons. In the case of SSC, where the seed photons come from synchrotron, the polarization degree after scattering is $\sim 40\%$; EC is unpolarized since the seed photons are unpolarized thermal photons. The above results have been confirmed by Krawczynski (2012) via Monte Carlo simulations. Based on the model in Bonometto et al. (1970), Zhang & Böttcher (2013) presents the first high-energy polarization prediction for LSP, ISP, and HSP blazars based on stationary blazar spectral fittings. As shown in Figure 1, the MeV polarization degree of the hadronic model is consistently higher than that of the leptonic model. While the X-ray polarization degree is also different for the leptonic and hadronic models, the X-ray emission may be contaminated by the low-energy synchrotron component, especially for the ISP and HSP blazars. Therefore, MeV polarimetry is ideal to distinguish the two models

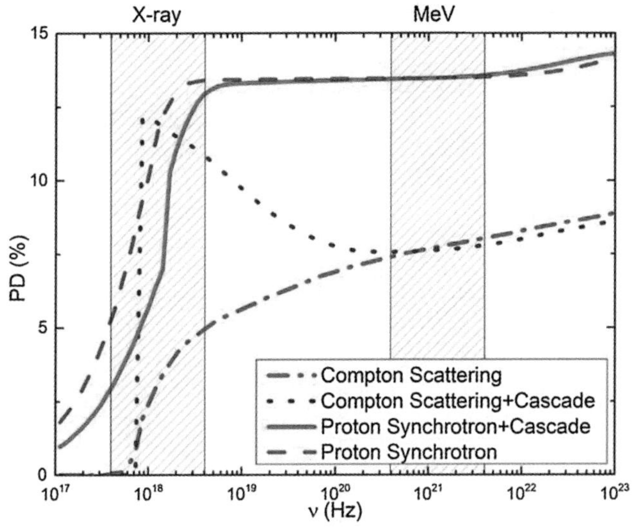

Figure 2. The X-ray to γ-ray spectral polarization degree based on the four leptonic and hadronic spectral fittings of the TXS 0506+056 neutrino event. The shaded regions correspond to the X-ray and MeV γ-ray bands. This figure is reproduced from Zhang et al. (2019).

for all blazar types. A key advantage of MeV polarimetry over the neutrino detection is that polarization is an intrinsic feature of photons, thus there is no ambiguity in whether the detected polarization is co-spatial with the emission.

Nonetheless, the above prediction is based on perfectly ordered magnetic fields. Optical polarization monitoring has matured in the last decade, showing typically $\lesssim 20\%$ polarization degree (Aliu et al. 2016; Blinov et al. 2018; Larionov et al. 2013; Marscher et al. 2010; Morozova et al. 2014). This implies a partially ordered magnetic field in the blazar emission region (Pushkarev et al. 2005; Zhang et al. 2015). If the low- and high-energy spectral components are co-spatial, the high-energy emission should suffer from the same depolarization effect due to partially ordered magnetic field as the low-energy emission. This depolarization factor can be estimated by the ratio of the observed optical polarization degree over the maximal synchrotron polarization degree in a perfectly ordered magnetic field by modeling the low-energy spectral index. However, many LSP blazars, especially flat spectrum radio quasars, have a strong thermal emission component in the optical band. In these sources, the optical polarization can be contaminated by the unpolarized thermal emission. Therefore, the depolarization factor should be calculated by first removing the thermal contamination. Paliya et al. (2018) has performed the above calculation and found that the hadronic model typically predicts comparable or higher MeV polarization degree than the optical band, while the leptonic model generally predicts less than half of the optical polarization. More recently, Zhang et al. (2019) have performed a detailed X-ray and MeV polarization prediction for the TXS 0506+056 flare and neutrino event. By comparing different spectral fitting models, this paper suggests that while X-ray and MeV polarization can both probe hadronic processes for TXS 0506+056, MeV polarimetry is unique in identifying the proton synchrotron process (see Figure 2. Since the proton synchrotron is only efficient in a strong magnetic field, preventing significant contributions from inverse Compton scattering, the γ-ray emission is dominated by the proton synchrotron. Spectral fittings then suggest that the observed γ-ray emission is produced by nonthermal protons extending to $\sim$EeV, i.e., the blazar emission region can

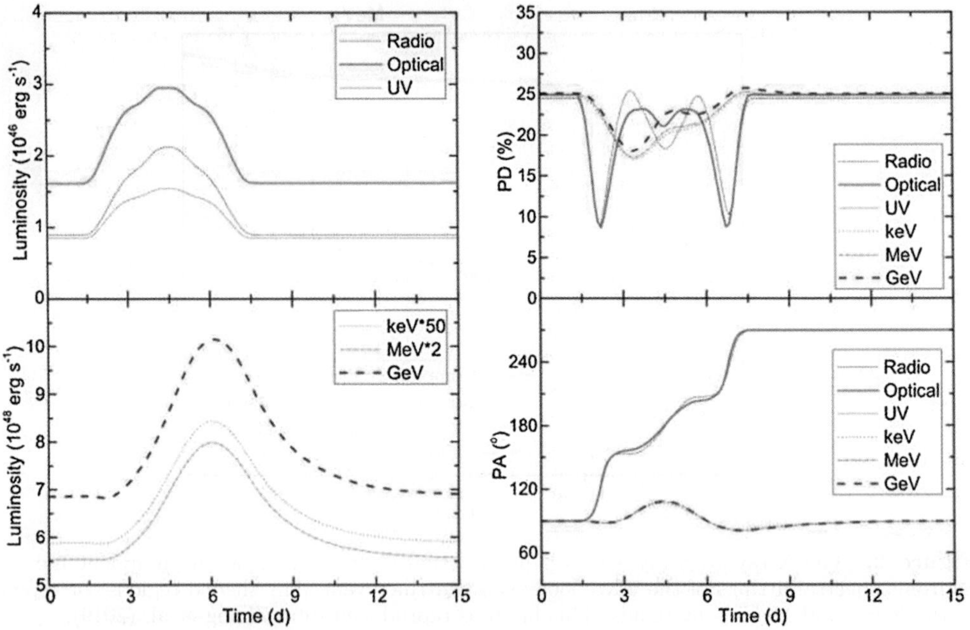

Figure 3. Light curves and polarization variations for the hadronic model. Left: multiwavelength light curves chosen at radio (30–300 GHz, navy short-dashed), optical (1.8–3.2 eV, thick orange solid), UV (3.3–6.2 eV, olive dot-dot-dashed), X-ray (60–200 keV, purple dotted), MeV γ-ray (5–200 MeV, AdEPT, pink dot-dashed), and GeV γ-ray (20 MeV–300 GeV, Fermi-LAT, thick violet dashed) bands. Due to the large bandwidth of the GeV light curve, it collects a much higher total luminosity than the keV and MeV bands. Hence, we manually boost those two bands by a fixed number to allow us to show them in the same figure. Right: multiwavelength PD and PA vs. time. Bands are chosen the same as light curves. This figure is reproduced from Zhang et al. (2016).

accelerate ultra-high-energy cosmic rays (UHECRs) (Böttcher et al. 2013; Cerruti et al. 2015). This study, although using rather simplified spectral fitting models, shows the potential that MeV polarimetry can explore the acceleration of UHECRs in blazars.

A major concern of detecting blazar MeV polarization is the variability. It is known that the blazar optical polarization can change, including large polarization angle swings (Blinov et al. 2015, 2018; Jorstad et al. 2022; Marscher et al. 2010). Since the MeV polarimeters cannot achieve the same temporal resolution as the optical counterparts due to the very low number of photons, they will require much longer time to integrate enough photons to determine the MeV polarization degree. If the polarization angle rotates significantly during this time window, the net polarization degree will drop. Zhang et al. (2016) has studied this issue with a multi-zone time-dependent simulation. Fortunately, the MeV polarization degree and angle are unlikely to change considerably during flares, due to the very slow proton cooling. In a typical blazar emission region, the proton cooling time scale is usually longer than the light crossing time scale, the latter is generally comparable to the time scale of magnetic field evolution either due to shocks or magnetic instabilities. On the other hand, electrons can quickly react to changes in the magnetic field, due to their very fast cooling. Zhang et al. (2016) considers a simulation setup with a major magnetic field morphology change in the blazar emission region, which drives a large polarization angle swing in the low-energy bands as shown in Figure 3. However, protons only slowly react to the changes due to their long cooling time, leading to a small dip in the polarization degree and minor fluctuations in the polarization angle,

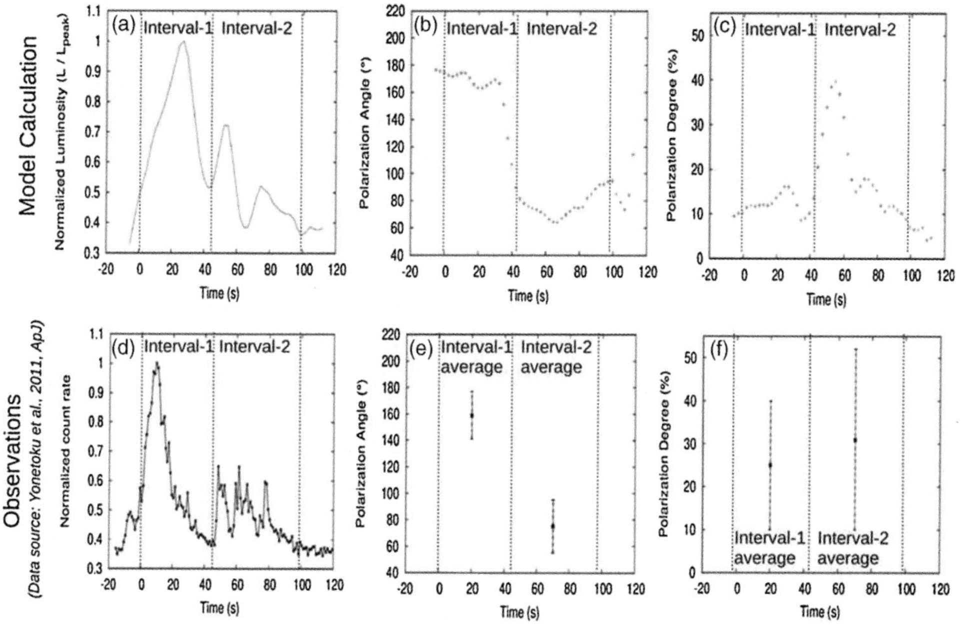

Figure 4. Comparison between integrated magnetohydrodynamic and polarized radiation transfer simulation results and GRB 100826A observation. Upper panels are the light curve as well as polarization angle and degree from the simulation, whereas the lower panels are the observation by Yonetoku et al. (2011). The light curves of both observation and calculation are normalized by their peak values. This figure is reproduced from Deng et al. (2016).

ideal for future MeV polarimeters to integrate a relatively long time for polarization detection. On the other hand, the leptonic scenario suffers from double depolarization due to partially ordered magnetic field. This is because the seed photons for SSC observe a partially ordered magnetic field thus get depolarized, while SSC from different parts of the emission region shall have their seed photons with different polarization status, resulting in a second depolarization. If the magnetic field can vary, the light crossing delay will cause the seed photons and the scattered photons arriving at different time, overall enhancing this double depolarization effect. This has been studied in Peirson et al. (2022), showing that the leptonic polarization degree is $\lesssim 30\%$ of the optical counterpart, almost impossible for future MeV polarimetry. Therefore, a detection of MeV polarization during blazar flares is almost certainly a signature of hadronic processes.

3. Understand GRB Prompt Emission with MeV Polarimetry

Similar to blazars, MeV polarimetry can also study the GRB prompt emission mechanisms. The synchrotron model clearly expects to have some level of γ-ray polarization, whether the magnetic field is ordered or disordered, although the former case predicts much higher polarization degree. Photons in the photospheric model experience scattering in the emission region before they can escape. However, unlike blazars, the scattering in GRBs is anisotropic, thus the resulting polarization degree can range from 0 to nearly 100%, depending on the viewing angle. Toma et al. (2009) presents a pioneering work on the GRB prompt polarization, in which they find an interesting distribution of prompt polarization degrees based on different models. This paper points to the potential of studying GRB prompt polarization statistically with a large number of accurate measurement to distinguish different models. A major issue here is that it is unlikely that all GRBs belong to one particular kind of prompt emission models. Current prompt

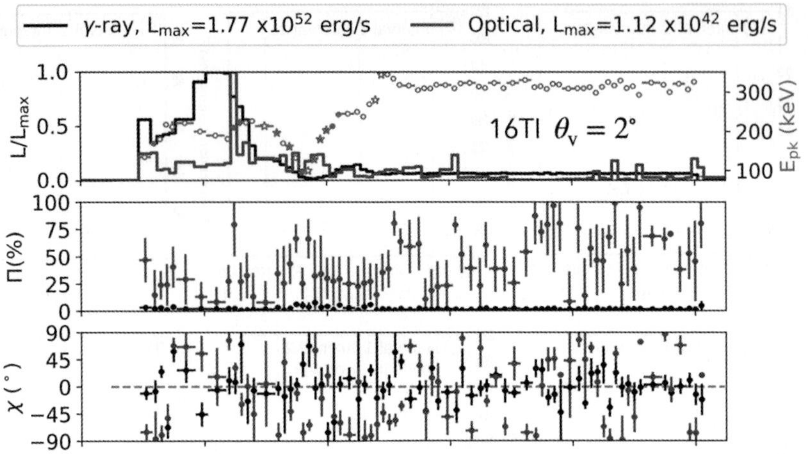

Figure 5. Light curve and polarization from integrated hydrodynamic and polarized radiation transfer simulations. The upper panel plots the time-resolved fitted spectral peak energy E_{pk} in green and the γ-ray and optical light curves in black and magenta respectively, where each light curve is normalized by its own maximum value. The middle panel plots the polarization degree of the γ-ray and optical photons in black and magenta. The lower panel shows the polarization angles of the mock observed γ-ray and optical bands. This figure is reproduced from Parsotan & Lazzati (2022).

polarization measurements, though with large error bars, do not show clear trends in the polarization degree distributions. Recently, Gill et al. (2020) present a more detailed study, but they still reach the conclusion that the synchrotron and photospheric models tend to have similar distributions of prompt polarization degrees. Their paper does argue that very high polarization ($\gtrsim 50\%$) is more naturally produced by the synchrotron model under typical jet configurations, but since the GRB jet morphology and physical conditions are not well understood either, collection of average prompt polarization degrees from many GRBs is insufficient to diagnose the GRB radiation mechanisms.

Temporally resolved prompt polarization degree and angle evolution are crucial to explore the GRB radiation. Based on the ICMART model, Deng et al. (2016) presents a pioneering numerical study of the prompt radiation and polarization signatures (see Figure 4. They employ integrated magnetohydrodynamic and polarized radiation transfer simulations to study the light curve and time-dependent polarization degree and angle from magnetic reconnection. This approach can self-consistently track the magnetic field structure and evolution so as to self-consistently derive polarization signatures. Very interestingly, they find a strong change of polarization degree and a $\sim 90°$ angle swing happening between two peaks in the prompt light curve, well resembling the observation presented in Yonetoku et al. (2011). We note that the polarization angle swing is not limited to $90°$ in the magnetic reconnection scenario. As demonstrated in recent integrated particle-in-cell and polarized radiation transfer simulations, reconnection can result in large angle swings that are accompanied by changes in polarization degree and flares (Zhang et al. 2021, 2022). Using a semi-analytical model, Gill & Granot (2021) draw similar conclusions in a recent paper, where they explore time-dependent polarization from more general synchrotron models with various magnetic field configurations (see Figure 5). For the photospheric model, Parsotan & Lazzati (2022) calculates the time-dependent polarization based on integrated hydrodynamic and polarized radiation transfer simulations, in which they also find a $90°$ polarization angle swing. Given the nature of photon scattering in the photospheric model, γ-ray polarization results from anisotropic scattering, thus the polarization degree and angle are closely related as shown in Parsotan & Lazzati (2022). Such correlation is not present in the synchrotron model. Additionally, the amplitude of continuous angle rotation is limited to 90° in the photospheric model, while the amplitude in a synchrotron model is arbitrary, at least based on a magnetic reconnection scenario.

4. Future Prospects

Future MeV polarimeters may examine the above theoretical predictions for blazars and GRBs. POLAR-2 is a planned γ-ray polarimeter for GRBs, which is projected to measure about 50 GRBs per year with equal or better quality compared to the best seen by POLAR (Hulsman 2020). Recently, NASA selected LEAP as a Mission of Opportunity, which is an MeV polarimeter that may be deployed on the International Space Station (McConnell et al. 2021). Both POLAR-2 and LEAP may have the capability to resolve time-dependent GRB polarization for very bright GRBs. However, they are not designed to observed persistent sources like blazars. COSI is a selected SMEX mission that can detect γ-ray polarization for both blazars and GRBs (Yang et al. 2018). Its sensitivity may be able to detect a couple of hadronic blazar polarization signatures during its mission span if they exhibit bright flares. AMEGO-X is a more powerful MeV polarimeter that is under development (Caputo et al. 2022). It may detect polarization from ten bright blazar flares per year if they are hadronic, more conclusively distinguishing the leptonic and hadronic blazar models. These missions can complement the ongoing X-ray polarimeter IXPE to cover the high-energy polarimetry for relativistic jets.

Theoretically, previous works have already demonstrated the importance of numerical simulations. With the development of more efficient numerical codes and connection between particle acceleration, fluid dynamics, and radiation transfer, future simulations are expected to link the kinetic scale of particles with the large-scale fluid dynamics to fully study radiation and polarization signatures under realistic blazar and GRB physical conditions. MeV polarimetry can become a key component of the multi-messenger astronomy in the next decade.

References

Abbott, B. P., Abbott, R., Abbott, T. D., et al. 2017, ApJ, 848, L13. doi:10.3847/2041-8213/aa920c

Abdo, A. A., Ackermann, M., Agudo, I., et al. 2010, ApJ, 716, 30. doi:10.1088/0004-637X/716/1/30

Abdollahi, S., Acero, F., Ackermann, M., et al. 2020, ApJS, 247, 33. doi:10.3847/1538-4365/ab6bcb

Aharonian, F. A. 2000, New Astron., 5, 377. doi:10.1016/S1384-1076(00)00039-7

Ajello, M., Angioni, R., Axelsson, M., et al. 2020, ApJ, 892, 105. doi:10.3847/1538-4357/ab791e

Aliu, E., Archambault, S., Archer, A., et al. 2016, A&A, 594, A76. doi:10.1051/0004-6361/201628744

Beloborodov, A. M. 2010, MNRAS, 407, 1033. doi:10.1111/j.1365-2966.2010.16770.x

Blinov, D., Pavlidou, V., Papadakis, I., et al. 2015, MNRAS, 453, 1669. doi:10.1093/mnras/stv1723

Blinov, D., Pavlidou, V., Papadakis, I., et al. 2018, MNRAS, 474, 1296. doi:10.1093/mnras/stx2786

Bonometto, S., Cazzola, P., & Saggion, A. 1970, A&A, 7, 292

Bonometto, S. & Saggion, A. 1973, A&A, 23, 9

Böttcher, M., Reimer, A., Sweeney, K., et al. 2013, ApJ, 768, 54. doi:10.1088/0004-637X/768/1/54

Böttcher, M. 2019, Galaxies, 7, 20. doi:10.3390/galaxies7010020

Caputo, R., Ajello, M., Kierans, C. A., et al. 2022, Journal of Astronomical Telescopes, Instruments, and Systems, 8, 044003. doi:10.1117/1.JATIS.8.4.044003

Cerruti, M., Zech, A., Boisson, C., et al. 2015, MNRAS, 448, 910. doi:10.1093/mnras/stu2691

Chen, X., Chatterjee, R., Zhang, H., et al. 2014, MNRAS, 441, 2188. doi:10.1093/mnras/stu713

Deng, W., Zhang, H., Zhang, B., et al. 2016, ApJ, 821, L12. doi:10.3847/2041-8205/821/1/L12

Dermer, C. D., Schlickeiser, R., & Mastichiadis, A. 1992, A&A, 256, L27

Diltz, C., Böttcher, M., & Fossati, G. 2015, ApJ, 802, 133. doi:10.1088/0004-637X/802/2/133

Ghisellini, G. & Madau, P. 1996, MNRAS, 280, 67. doi:10.1093/mnras/280.1.67

Giannios, D. & Uzdensky, D. A. 2019, MNRAS, 484, 1378. doi:10.1093/mnras/stz082

Gill, R., Granot, J., & Kumar, P. 2020, MNRAS, 491, 3343. doi:10.1093/mnras/stz2976

Gill, R. & Granot, J. 2021, MNRAS, 504, 1939. doi:10.1093/mnras/stab1013

Goldstein, A., Veres, P., Burns, E., et al. 2017, ApJ, 848, L14. doi:10.3847/2041-8213/aa8f41

Hjorth, J., Sollerman, J., Møller, P., et al. 2003, Nature, 423, 847. doi:10.1038/nature01750

Hulsman, J. 2020, Proc. SPIE, 11444, 114442V. doi:10.1117/12.2559374

IceCube Collaboration, Aartsen, M. G., Ackermann, M., et al. 2018, Science, 361, eaat1378. doi:10.1126/science.aat1378

Jorstad, S. G., Marscher, A. P., Raiteri, C. M., et al. 2022, Nature, 609, 265. doi:10.1038/s41586-022-05038-9

Krawczynski, H. 2012, ApJ, 744, 30. doi:10.1088/0004-637X/744/1/30

Larionov, V. M., Jorstad, S. G., Marscher, A. P., et al. 2013, ApJ, 768, 40. doi:10.1088/0004-637X/768/1/40

Li, H. & Kusunose, M. 2000, ApJ, 536, 729. doi:10.1086/308960

MacFadyen, A. I., Woosley, S. E., & Heger, A. 2001, ApJ, 550, 410. doi:10.1086/319698

Mannheim, K. 1993, A&A, 269, 67. doi:10.48550/arXiv.astro-ph/9302006

Maraschi, L., Ghisellini, G., & Celotti, A. 1992, ApJ, 397, L5. doi:10.1086/186531

Marscher, A. P. & Gear, W. K. 1985, ApJ, 298, 114. doi:10.1086/163592

Marscher, A. P., Jorstad, S. G., Larionov, V. M., et al. 2010, ApJ, 710, L126. doi:10.1088/2041-8205/710/2/L126

Morozova, D. A., Larionov, V. M., Troitsky, I. S., et al. 2014, AJ, 148, 42. doi:10.1088/0004-6256/148/3/42

McConnell, M. L., Baring, M., Bloser, P., et al. 2021, Proc. SPIE, 11821, 118210P. doi:10.1117/12.2594737

Mücke, A., Protheroe, R. J., Engel, R., et al. 2003, Astroparticle Physics, 18, 593. doi:10.1016/S0927-6505(02)00185-8

Paliya, V. S., Zhang, H., Böttcher, M., et al. 2018, ApJ, 863, 98. doi:10.3847/1538-4357/aad1f0

Parsotan, T. & Lazzati, D. 2022, ApJ, 926, 104. doi:10.3847/1538-4357/ac4093

Pe'er, A., Mészáros, P., & Rees, M. J. 2006, ApJ, 642, 995. doi:10.1086/501424

Peirson, A. L., Liodakis, I., & Romani, R. W. 2022, ApJ, 931, 59. doi:10.3847/1538-4357/ac6a54

Petropoulou, M., Dimitrakoudis, S., Padovani, P., et al. 2015, MNRAS, 448, 2412. doi:10.1093/mnras/stv179

Pushkarev, A. B., Gabuzda, D. C., Vetukhnovskaya, Y. N., et al. 2005, MNRAS, 356, 859. doi:10.1111/j.1365-2966.2004.08535.x

Rees, M. J. & Meszaros, P. 1994, ApJ, 430, L93. doi:10.1086/187446

Rees, M. J. & Mészáros, P. 2005, ApJ, 628, 847. doi:10.1086/430818

Rybicki, G. B. & Lightman, A. P. 1986, Radiative Processes in Astrophysics, by George B. Rybicki, Alan P. Lightman, pp. 400. ISBN 0-471-82759-2. Wiley-VCH, June 1986., 400

Sahu, S., Oliveros, A. F. O., & Sanabria, J. C. 2013, Phys. Rev. D, 87, 103015. doi:10.1103/PhysRevD.87.103015

Sikora, M., Begelman, M. C., & Rees, M. J. 1994, ApJ, 421, 153. doi:10.1086/173633

Toma, K., Sakamoto, T., Zhang, B., et al. 2009, ApJ, 698, 1042. doi:10.1088/0004-637X/698/2/1042

Yang, C.-Y., Lowell, A., Zoglauer, A., et al. 2018, Proc. SPIE, 10699, 106992K. doi:10.1117/12.2312556

Yonetoku, D., Murakami, T., Gunji, S., et al. 2011, ApJ, 743, L30. doi:10.1088/2041-8205/743/2/L30

Zhang, B. & Yan, H. 2011, ApJ, 726, 90. doi:10.1088/0004-637X/726/2/90

Zhang, H. & Böttcher, M. 2013, ApJ, 774, 18. doi:10.1088/0004-637X/774/1/18

Zhang, H., Chen, X., Böttcher, M., et al. 2015, ApJ, 804, 58. doi:10.1088/0004-637X/804/1/58

Zhang, H., Diltz, C., & Böttcher, M. 2016, ApJ, 829, 69. doi:10.3847/0004-637X/829/2/69

Zhang, H., Fang, K., Li, H., et al. 2019, ApJ, 876, 109. doi:10.3847/1538-4357/ab158d

Zhang, H. 2019, Galaxies, 7, 85. doi:10.3390/galaxies7040085

Zhang, H., Li, X., Giannios, D., et al. 2021, ApJ, 912, 129. doi:10.3847/1538-4357/abf2be

Zhang, H., Li, X., Giannios, D., et al. 2022, ApJ, 924, 90. doi:10.3847/1538-4357/ac3669

The Multimessenger Chakra of Blazar Jets
Proceedings IAU Symposium No. 375, 2023
I. Liodakis, M. F. Aller, H. Krawczynski, A. Lähteenmäki & T. J. Pearson, eds.
doi:10.1017/S1743921323001151

J2102+6015: a potential distant multimessenger?

Leonid I. Gurvits[1,2], Sándor Frey[3,4], Máté Krezinger[5,3], Oleg Titov[6], Tao An[7], Yingkang Zhang[7], Alexander G. Polnarev[8], Krisztina É. Gabányi[5,9,3], Krisztina Perger[3] and Alexey Melnikov[10]

[1]Joint Institute for VLBI ERIC, Oude Hoogeveensedijk 4, 7991 PD Dwingeloo, The Netherlands

[2]Aerospace Faculty, Delft University of Technology, Kluyverweg 1, 2629 HS Delft, The Netherlands

[3]Konkoly Observatory, ELKH Research Centre for Astronomy and Earth Sciences (MTA Centre of Excellence), Konkoly Thege Miklós út 15-17, H-1121, Budapest, Hungary

[4]Institute of Physics and Astronomy, ELTE Eötvös Loránd University, Pázmány Péter sétány 1/A, H-1117, Budapest, Hungary

[5]Department of Astronomy, Institute of Physics and Astronomy, ELTE Eötvös Loránd University, Pázmány Péter sétány 1/A, H-1117, Budapest, Hungary

[6]Geoscience Australia, PO Box 378, Canberra 2601, Australia

[7]Shanghai Astronomical Observatory, CAS, 80 Nandan Road, Shanghai 200030, China

[8]Queen Mary University of London, London E1 4NS, United Kingdom

[9]ELKH-ELTE Extragalactic Astrophysics Research Group, Eötvös Loránd University, Pázmány Péter sétány 1/A, H-1117, Budapest, Hungary

[10]Institute of Applied Astronomy, Russian Academy of Sciences, Kutuzova Embankment 10, St. Petersburg 191187, Russia

Abstract. We present and briefly discuss results of several studies of the source J2102+6015 with tentatively defined redshift $z = 4.575$ which demonstrates unusual properties in imaging and astrometric VLBI observations. Its properties might be considered as indications on the supermassive black hole binary which can be considered as a so far rare example of a high-redshift source of known electromagnetic and, possibly, predictable gravitational wave emissions.

Keywords. SMBH binary, AGN, VLBI, multimessenger

1. Introduction

Active galactic nuclei (AGN) are the most powerful "engines" in the Universe. Due to their enormous luminosity they are visible and can be studied with modern astronomy instruments at any cosmological distances. Owing to the progress of optical telescopes, including spaceborne ones, as the JWST, the redshift frontier of known AGN currently exceeds $z = 7$ corresponding to the Universe age of 5% of the present value. The high-redshift objects at this frontier, just as their low-redshift counterparts, appear to harbour and be "energised" by supermassive black holes (SMBH). The masses of the latter reach and even exceed $10^9 M_{\odot}$. In many observable manifestations, the high-redshift AGN appear to be very similar to their low-redshift counterparts. This similarity raises an important question: what are physical processes for which a relatively short time on cosmological scales, hundreds of millions of years or shorter after Big Bang, is sufficient to assemble SMBHs. This topical question is being addressed by many observational techniques in all domains of the electromagnetic spectrum. In radio domain, high angular resolution studies with very long baseline interferometry (VLBI) provide insight into physical processes at the linear scales from hundreds of parsecs down to sub-parsecs throughout the entire range of redshifts. At the high-redshift range, to date these studies have encompassed more than 70 AGN at $z > 4$ (Krezinger et al. 2022; Gabányi et al. 2023; Perger et al. 2023 and references therein).

Recently a small sample of four $z > 4.3$ quasars was observed with several VLBI networks in the so called astrometric mode aiming to detect systematic absolute proper motion of cosmological origin (Titov et al. 2023). While this study did not produce a conclusive evidence of any anomalies of cosmological origin in the positional variations of these sources, one of them, J2102+6015, appeared to deserve attention in view of its potential significance for future multimessenger studies.

2. An enigmatic object J2102+6015

The source J2102+6015 (2101+600) was identified as a quasar at $z = 4.5749$; this value was noted by the authors as somewhat unreliable due to a low signal-to-noise ratio of the details in the optical spectrum (Sowards-Emmerd et al. 2004). No other measurements of the source's redshift are known to date. The source was observed with various VLBI networks at 2.3, 4.9, 8.4 and 22 GHz (Frey et al. 2018; Zhang et al. 2021; Frey et al. 2023; Titov et al. 2023). These observations resulted in the unusual findings requiring further in-depth investigations.

All available to date imaging VLBI results confirm that at all four frequency bands the source's morphology is dominated by two major features separated by about 10 mas in EW direction (see insets in Fig. 1). Frey et al. (2023) report new European VLBI Network (EVN) observations at 5 and 22 GHz, and in combination with earlier publications on VLBI results at 2.3 and 8.4 GHz, discuss a set of VLBI images obtained over the period 1994–2021. The eastern feature is stronger and brighter at all frequencies at all epochs. Both features can be model-fitted by two or three Gaussian components. The angular separation between major components of the eastern and western features over the period 2006–2019 is reported to grow with the rate 0.023 ± 0.011 mas yr^{-1}, corresponding to the apparent transversal velocity of $(2.8 \pm 1.4)c$ (Zhang et al. 2022). However, to date, one and only one observing epoch resulted in a triple structure in J2101+6015, in which the known double morphology is amended by a third component located along the same east-west line between the eastern and western features. This is the Very Long Baseline Array (VLBA) observation BZ064 at 8.4 GHz conducted on 2017-02-05. Zhang et al. (2021) and Frey et al. (2023) consider arguments for and against classifying the source as a compact symmetric object. Such a classification would be well fitted by the existence of a weak central component associated with the nucleus of the source. However, barring a technical problem in the imaging experiment BZ064 responsible for the single-case erroneous appearance of the central component, a potential variability in the strength and brightness of structural features of the source might have meaningful implications for the model of this object.

A further intriguing property of the source J2102+6015 is reported by Titov et al. (2023) on the basis of astrometric monitoring over the period 2017–2021 at 2.3 and 8.4 GHz. Using a standard for VLBI astrometry method of measuring absolute celestial position of the source, the authors presented apparently quasi-periodic variations of the source's right ascension and declination with the period about 3 years (Fig. 7 in Titov et al. 2023) and amplitudes in both coordinates of about 0.8 mas. These variations, if placed on the celestial sphere, produce a quasi-spiral pattern, reproduced in

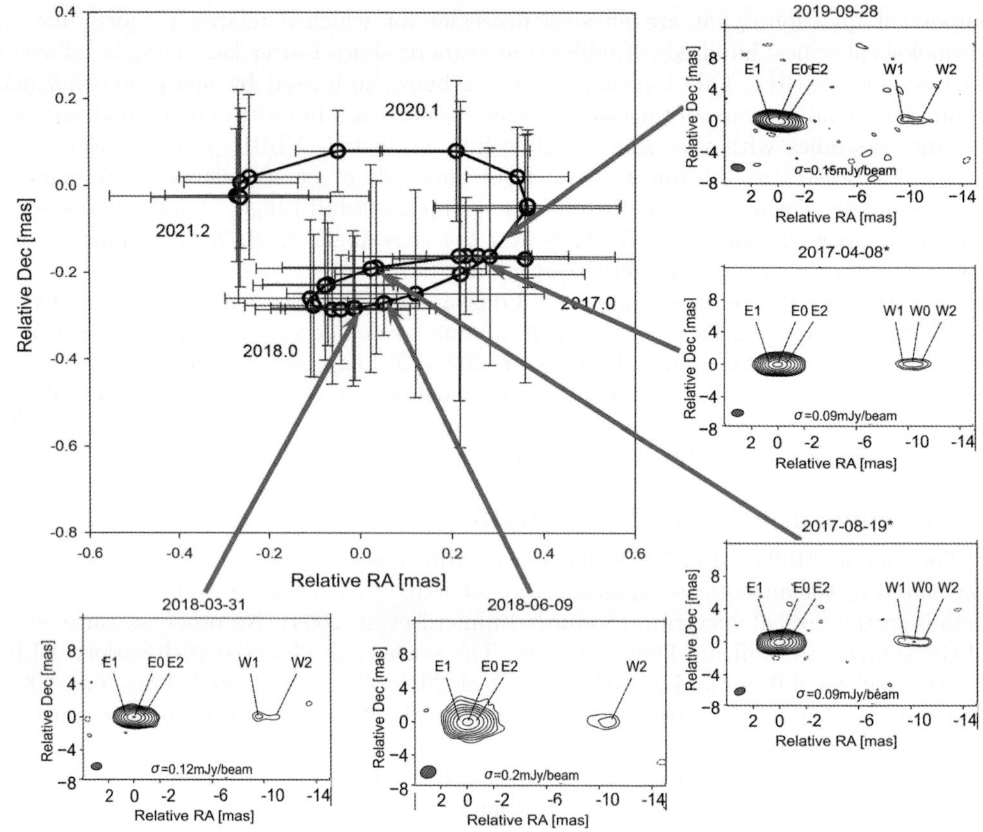

Figure 1. A celestial track of the absolute positions of the source J2102+6015 in the period 2017.0 to 2021.2 determined on the basis of VLBI measurements, adopted from Titov et al. (2023). The error bars show $\pm 1\sigma$ formal uncertainty. The insets show the morphology of the source with the total flux density between ~ 150 mJy and ~ 200 mJy in all epochs obtained in imaging VLBI observations at 8.4 GHz, adopted from Zhang et al. (2021). The elliptical Gaussian restoring beam is shown in the lower left corner of each inset. The lowest contours are at $\pm 3\sigma$ of the image noise shown in each inset, positive contour levels increase by a factor of 2. The images attributed to 2017-04-08 and 2017-08-19 were obtained from VLBI data stacked over several close dates; see Zhang et al. (2021) for details.

Fig. 1. This pattern enabled the authors to consider a "toy model" of the source encompassing a supermassive black hole binary (SMBHB). Indeed, an orbital motion of the binary components might produce apparent oscillations of the celestial coordinates of the source. We note that such an interpretation of the apparent spiral pattern shown in Fig. 1 is not unique. However, given the importance of astrophysical and cosmological implications of SMBHB, the model warrants further deliberations and possible verifiable observational manifestations.

3. J2102+6015, a potential high-redshift multi-messenger

The "toy model" of the source mentioned above (see Titov et al. 2023) offers a simple estimate of the semi-major axis of the suspected binary system

$$a = 0.32 \times 10^{-2} \times \left(\frac{M}{10^9 \times M_{\odot}}\right)^{1/3} \times \left(\frac{T \times (1+z)}{3.0\,\mathrm{yr}}\right)^{2/3} \quad \mathrm{pc}, \qquad (1)$$

where a is the semi-major axis of the elliptical orbit of the binary system of the total mass M (normalised here to the benchmark value of $10^9 M_\odot$), T is its observer-frame period (in years). Assuming equal masses of $0.5 \times 10^9 M_\odot$ for each of the two components for purely illustrative reasons, the semi-major axis of the system would only be $a \simeq 0.25 \times 10^{-2}$ pc, or $\simeq 50$ times the gravitational radius of the SMBHB component, $r_\mathrm{g} = 2G(M/2)c^{-2} \approx 0.48 \times 10^{-4}$ pc.

A potential binary nature of the source J2102+6015 has also been suggested by Zhang et al. (2021). Indeed, flat radio spectra of the eastern and western morphology features separated by the projected distance of ~ 70 pc make them plausible candidates for typical AGN cores. However, their association with components of a suspected SMBHB is inconsistent with the dynamical "toy model" described by equation (1): the total mass of such a binary system would be $\sim 10^{16} M_\odot$ even without accounting for the projection effect. This is a very unrealistic value. For remaining within the framework of SMBHB nature of the observed oscillation of the absolute celestial position of the source, another binary configuration should be found. The enigmatic central morphological component, reported by Zhang et al. (2021) might help to find such a configuration. However, given that the astrometric behaviour illustrated in Fig. 1 is defined to great extent by the brightest eastern morphological feature, a suspected binary system, responsible for the astrometric oscillations, should be embedded within that eastern feature.

Irrespective of the specific parameters of the SMBHB model and the source's redshift, the source J2102+6015 might be considered as a generator of gravitational waves and a likely contributor into the global gravitational wave background. SMBHB systems evolve inevitably due to the interaction of a less massive object with an accretion disk around a more massive black hole. In this case, the dissipation of the orbital energy occurs, as a result of which the two black holes approach each other (see, for example, Ivanov et al. 1999). In the process of decreasing the size of the binary's orbit, the role of gravitational waves increases. Gravitational waves carry away the energy and angular momentum of the binary system (see, for example, Ivanov et al. 2015). How does the radiation of gravitational waves affect the observed parameters of binary systems that are at later stages of evolution than the system considered above, whether such gravitational waves can be detected in the foreseeable future? The source J2102+6015 presented here might become an efficient testbed for finding the answers.

Acknowledgements

LIG is grateful to the Leids Kerkhoven-Bosscha Fonds for supporting his attendance at the IAU Symposium No. 375 (grant 22.2.028). SF, KEÅLG and KP were supported by the Hungarian National Research, Development and Innovation Office (OTKA K134213).

Note added in proof

A recent re-evaluation of the available optical spectrum of the source J2102+6015 provides indications that its redshift might be $z = 1.24$ rather than $z = 4.575$. However, this does not change our view at this source as a potentially highly attractive multimessenger.

References

Frey, S., An, T., Gabányi, K., Gurvits, L., Krezinger, M., Melnikov, A., Mohan, P., Paragi, Z., Perger, K., Shu, F., Titov, O., de Vicente, P., & Zhang, Y. 2023, J2102+6015: an Intriguing Radio-loud Active Galactic Nucleus in the Early Universe. *PoS*, EVN2022(022), arXiv.2301.07355.

Frey, S., Titov, O., Melnikov, A. E., de Vicente, P., & Shu, F. 2018, High-resolution radio imaging of two luminous quasars beyond redshift 4.5. *A&A*, 618, A68.

Gabányi, K. É., Belladitta, S., Frey, S., Orosz, G., Gurvits, L. I., Rozgonyi, K., An, T., Cao, H., Paragi, Z., & Perger, K. 2023, Very long baseline interferometry observations of the high-redshift blazar candidate J0141-5427. *PASA*, 40, e004.

Ivanov, P. B., Papaloizou, J. C. B., Paardekooper, S. J., & Polnarev, A. G. 2015, The evolution of a binary in a retrograde circular orbit embedded in an accretion disk. *A&A*, 576, A29.

Ivanov, P. B., Papaloizou, J. C. B., & Polnarev, A. G. 1999, The evolution of a supermassive binary caused by an accretion disc. *MNRAS*, 307(1), 79–90.

Krezinger, M., Perger, K., Gabányi, K. É., Frey, S., Gurvits, L. I., Paragi, Z., An, T., Zhang, Y., Cao, H., & Sbarrato, T. 2022, Radio-loud Quasars above Redshift 4: Very Long Baseline Interferometry (VLBI) Imaging of an Extended Sample. *ApJS*, 260(2), 49.

Perger, K., Zhang, Y., Frey, S., An, T., Gabányi, K., Gurvits, L., Hwang, C.-Y., Koptelova, E., Paragi, Z., & Wang, A. 2023, High-resolution Imaging of Two Radio Quasars at the End of the Reionization Epoch. *PoS*, EVN2022(024), arXiv:2301.09947.

Sowards-Emmerd, D., Romani, R. W., Michelson, P. F., & Ulvestad, J. S. 2004, Blazar Counterparts for 3EG Sources at $-40^\circ <$Decl.$< 0^\circ$: Pushing South through the Bulge. *ApJ*, 609(2), 564–575.

Titov, O., Frey, S., Melnikov, A., Shu, F., Xia, B., Gonzalez, J., Tercero, B., Gurvits, L. I., de Witt, A., McCallum, J., Kharinov, M., Zimovsky, V., & Krezinger, M. 2023, Astrometric Apparent Motion of High-Redshift Radio Sources. *AJ*, 165(69), 12.

Zhang, Y., An, T., Frey, S., Gabányi, K. É., & Sotnikova, Y. 2022, Radio Jet Proper-motion Analysis of Nine Distant Quasars above Redshift 3.5. *ApJ*, 937(1), 19.

Zhang, Y., An, T., Frey, S., Yang, X., Krezinger, M., Titov, O., Melnikov, A., de Vicente, P., Shu, F., & Wang, A. 2021, J2102+6015: a young radio source at $z = 4.575$. *MNRAS*, 507(3), 3736–3744.

The Multimessenger Chakra of Blazar Jets
Proceedings IAU Symposium No. 375, 2023
I. Liodakis, M. F. Aller, H. Krawczynski, A. Lähteenmäki & T. J. Pearson, eds.
doi:10.1017/S1743921323000893

VLBI Scrutiny of a New Neutrino-Blazar Multiwavelength-Flare Coincidence

F. Eppel[1], M. Kadler[1], E. Ros[2], F. Rösch[1], J. Heßdörfer[1], P. Benke[2], P. G. Edwards[3], C. M. Fromm[1], M. Giroletti[4], A. Gokus[5], J. L. Gómez[6], S. Hämmerich[7], D. Kirchner[1], Y. Y. Kovalev[2,8], T. P. Krichbaum[2], M. L. Lister[9], C. Nanci[4], R. Ojha[10], G. F. Paraschos[2], A. Plavin[8], A. C. S. Readhead[11], J. Stevens[3] and P. Weber[1]

[1]Julius-Maximilians-Universität Würzburg, Physikalisches Institut, Lehrstuhl für Astronomie, Emil-Fischer-Straße 31, 97074 Würzburg, Germany

[2]Max-Planck-Institut für Radioastronomie, Auf dem Hügel 69, 53121 Bonn, Germany

[3]CSIRO Astronomy and Space Science, ATNF, PO Box 76. Epping NSW 1710. Australia

[4]INAF, Istituto di Radio Astronomia, Via Piero Gobetti, 41029 Bologna, Italy

[5]McDonnell Center for Space Science, Washington University, MSC 1105-109-03, One Brookings Drive, St. Louis, MO 63130-4899, USA

[6]Instituto de Astrofísica de Andalucía, Gta. de la Astronomia, 18008 Granada, Spain

[7]Remeis Observatory and Erlangen Centre for Astroparticle Physics, Universität Erlangen-Nürnberg, Sternwartstr. 7, 96049 Bamberg, Germany

[8]Lebedev Physical Institute, Leninskii pr. 53, 119991, Moscow, Russia

[9]Department of Physics and Astronomy, Purdue University, 525 Northwestern Avenue, West Lafayette, IN 47907, USA

[10]NASA, Goddard Space Flight Center, Astrophysics Science Division, Code 661, Greenbelt, MD 20071, USA

[11]California Institute of Technology, 1200 E California Blvd, Pasadena, CA 91125, USA

Abstract. In recent years, evidence has started piling up that some high-energy cosmic neutrinos can be associated with blazars in flaring states. On 2022 February 26, a new blazar-neutrino coincidence was reported: the track-like neutrino event IC220225A detected by IceCube is spatially coincident with the flat-spectrum radio quasar PKS 0215+015. Like previous associations, this source was found to be in a high optical and γ-ray state. Moreover, the source showed a bright radio outburst, which substantially increases the probability of a true physical association. We have performed six observations with the VLBA shortly after the neutrino event with a monthly cadence and are monitoring the source with the Effelsberg 100m-Telescope, and with the Australia Compact Telescope Array. Here, we present first results on the contemporary parsec-scale jet structure of PKS 0215+015 in total intensity and polarization to constrain possible physical processes leading to neutrino emission in blazars.

Keywords. AGN, Blazar, Neutrino, Multimessenger, VLBI

1. Introduction

Gamma-ray blazars have long been predicted to be bright enough in neutrino emission and to exhibit a neutrino spectrum flat enough to exceed the atmospheric background at sufficiently high energies above about 100 TeV (e.g., Mannheim *et al.* 1992). The fundamental assumption in such calculations is that both neutrinos and γ-ray photons are produced by the same accelerated protons in the jet (hadronic processes as the dominant origin for the γ-ray emission of blazars). Kadler *et al.* (2016) have shown that within this general framework the long-term average γ-ray emission of blazars as a class is indeed in agreement with both the measured all-sky flux of very-high-energy neutrinos, and the spectral slope of the IceCube signal. However, the association of an individual blazar with a single neutrino has been limited by the poor positional accuracy of the shower-like events considered in that study. A major step forward was made when the track-like muon neutrino event IC 170922A was found to coincide with the γ-ray blazar TXS 0506+056 (GCN Circular 21916). Thanks to the much improved positional accuracy of the reconstruction for that event and the variability of the γ-ray emission, the association between IC 170922A and TXS 0506+056 was found to be on the level of $3\,\sigma$ (IceCube Collaboration *et al.* 2018). The general statistical correlation between radio-bright AGN samples and IceCube neutrinos that was revealed only recently (Plavin *et al.* 2020, 2021; Hovatta *et al.* 2021) is a fundamental independent observational breakthrough that strongly calls for high angular-resolution radio data of individual coincident neutrino detections and radio flares to solve this important question (e.g., Nanci *et al.* 2022). On 2022 February 26, the IceCube collaboration announced a new high-energy neutrino alert (IC220225A†) which is spatially coincident with the flat-spectrum radio quasar (FSRQ) PKS 0215+015. Like previous neutrino-blazar associations this source was found to be in a high-optical and γ-ray state (Garrappa *et al.* 2022; Nesci 2022). Members of our team found the source to be in a bright radio outburst (Kadler *et al.* 2022; Plavin *et al.* 2022), which increases the probability of a true physical association. In the following we discuss the first results of our ongoing follow-up observing campaign of the event with the Australia Telescope Compact Array (ATCA), the Effelsberg 100-m telescope and the Very Long Baseline Array (VLBA).

2. Observations & Analysis

ATCA has monitored PKS 0215+015 at multiple frequencies ranging from 2.1 GHz to 50 GHz since 2004 as part of calibrator and AGN monitoring programs (Stevens *et al.* 2012). The data are being processed by a standard pipeline and are made publicly available‡. Moreover, we added the source to the regular monitoring sample of the TELAMON program (Kadler *et al.* 2021) shortly after the neutrino detection (first observation on 2022 Feb 27). TELAMON observations are carried out with the Effelsberg 100-m telescope. Flux densities are derived by comparison with suitable secondary calibrators and averaged over three different sub-bands (20 mm, 14 mm and 7 mm). A detailed description of these observations will be published elsewhere. Triggered by the intriguing association of the neutrino event with the flare, we have successfully requested VLBA target-of-opportunity observations of PKS 0215+015 for six different sessions distributed over five months after the neutrino event (~ one per month) through a director's discretional time (DDT) proposal. These observations were carried out at 2 cm, 1.3 cm and 7 mm including full polarization information for every band. In addition to this DDT request the source has been observed with the VLBA at 2 cm as part of the MOJAVE program in the past

† `https://gcn.gsfc.nasa.gov/gcn/gcn3/31650.gcn3`

‡ `https://www.narrabri.atnf.csiro.au/calibrators/calibrator_database_viewcal.html?source=0215+015&detailed=true`

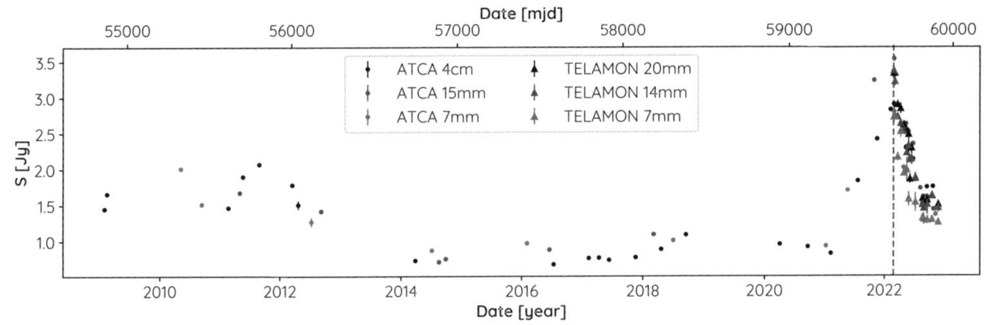

Figure 1. Radio Light Curve of PKS 0215+015 from TELAMON (Effelsberg) and ATCA observations since 2009. The source exhibits a major radio flare coincident with neutrino alert IC220225A (dashed red line). This radio flare is also coincident with a γ-ray flare observed by *Fermi-LAT*

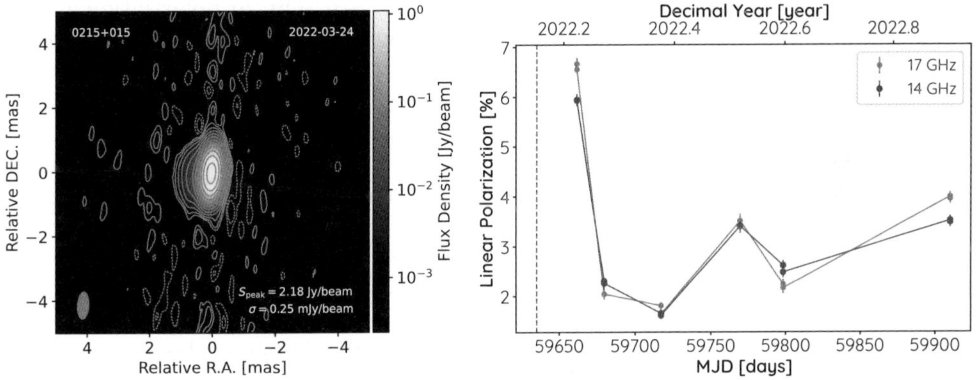

Figure 2. *Left:* VLBI-Image of PKS 0215+015 in total intensity at a wavelength of 1.3 cm (23 GHz) observed with the VLBA on 2022 March 24, one month after the neutrino-detection. *Right:* Linear polarization fraction of the same source as observed with the Effelsberg 100-m telescope as part of the TELAMON program. The dashed red line indicates the time of the neutrino event.

(Lister *et al.* 2019) and was reinstated into the regular monitoring just shortly before the neutrino event.

3. First Results

3.1. *Radio-Lightcurve*

Figure 1 shows the radio light curve of PKS 0215+015 compiled from selected ATCA data since 2009 and TELAMON observations. The source clearly reached its historical maximum near the time of the neutrino event and has been decreasing in radio brightness since. Publicly available *Fermi-LAT* data indicate a simultaneous γ-ray outburst†.

3.2. *VLBI-Structure*

The VLBA data were processed in a standard way using AIPS (Greisen 1990) for delay and amplitude calibration and DIFMAP (Shepherd 1997) for self-calibration and imaging. Here, we focus on the data from the first observing epoch on 2022 March 24. Figure 2 (left) shows the first image of PKS 0215+015 after the neutrino event at

† `https://fermi.gsfc.nasa.gov/ssc/data/access/lat/LightCurveRepository/source.php?source_name=4FGL_J0217.8+0144`

23 GHz from this epoch. The source shows a very compact structure with a jet pointed in the Eastern direction with one visible jet component. This jet component is also visible in the 2 cm and 7 mm images from 2022 March 24, and partially blended with the core in all three wavebands. We have estimated the brightness temperature T_B of the core component using the method described by Kovalev *et al.* (2005). At 2 cm and 1.3 cm, we find $T_B \sim 2.2 \cdot 10^{13}$ K, while at 7 mm $T_B \sim 3.5 \cdot 10^{12}$ K. We note that all derived brightness temperatures exceed the equipartition limit (Readhead 1994) and therefore suggest Doppler beaming.

3.3. *Polarization Properties*

From the Effelsberg observations, we derived polarization properties, namely linear polarization and electric vector position angle (EVPA) of the radio emission. Figure 2 (right) shows preliminary results of this analysis, indicating that the source was in an elevated polarization state close to the neutrino event ($\sim 6\,\%$ linear polarization at 2 cm) which decayed afterwards. The EVPA does not show any significant changes. Moreover, we have performed a preliminary polarization calibration of the first VLBA epoch using PolSolve (Marti-Vidal *et al.* 2021) in CASA which suggests that the polarized emission is core-dominated, just like the emission in total intensity.

4. Conclusions & Outlook

Just like three other blazars that were previously associated with neutrino events (PKS 1424−418, TXS 0506+056 and PKS 1502+106, Kadler *et al.* 2016; Ros *et al.* 2020; Karamanavis *et al.* 2015), PKS 0215+015 exhibits a very compact core-jet morphology. This suggests that neutrino production in PKS 0215+015 happens close to the central core. In the case of TXS 0506+056 and PKS 1424−418, the neutrino-associated radio flare also happened mostly in the compact core region. In TXS 0506+056, the core flux density increase was accompanied by a core expansion with apparent superluminal velocity (Ros *et al.* 2020). Moreover, a limb-brightened structure, as predicted by Tavecchio *et al.* (2015), was found in TXS 0506+056, PKS 1424−418 and PKS 1502+106. The similar sub-parsec scale structures of these possible neutrino-blazar associations could therefore indicate that neutrino-production in blazars usually happens in similar environments and scales. In order to further investigate this, we will search the remaining five VLBA epochs of PKS 0215+015 for these signatures. We will investigate the kinematics of the jet component to find out whether its ejection happened close in time to the neutrino event. Due to its compactness it can be challenging to detect a limb-brightening in the jet, however the collected polarization data (i.e., the orientation of the EVPA) can equally provide evidence for the existence of a "spine-sheath" structure.

Acknowledgements

This work is based on observations with the 100-m telescope of the MPIfR (Max-Planck-Institut für Radioastronomie) at Effelsberg. FE, FR, JH, and MK acknowledge support from the Deutsche Forschungsgemeinschaft (DFG, grants 447572188, 434448349, 465409577). We acknowledge the M2FINDERS project from the European Research Council (ERC) under the European Union's Horizon 2020 research and innovation programme (grant agreement No 101018682). This research has made use of data from the MOJAVE database that is maintained by the MOJAVE team (Lister *et al.* 2019). The National Radio Astronomy Observatory is a facility of the National Science Foundation operated under cooperative agreement by Associated Universities, Inc.

References

S. Garrappa *et al.* 2022, The Astronomer's Telegram, 15243
E. W. Greisen 1990, Acquisition, Processing and Archiving of Astronomical Images, 125–142
T. Hovatta *et al.* 2021, *A&A*, 650, A83
IceCube Collaboration *et al.* 2018, Science, 361, 1378
M. Kadler *et al.* 2016, NatPhys, 12, 807
M. Kadler *et al.* 2021, 37th International Cosmic Ray Conference
M. Kadler *et al.* 2022, The Astronomer's Telegram, 15245
V. Karamanavis *et al.* 2015, *A&A*, 586, A60
Y. Y. Kovalev *et al.* 2005, *AJ*, 130, 2473
M. Lister *et al.* 2019, *ApJ*,874, 43
K. Mannheim *et al.* 1992, *A&A*, 260, L1
I. Martí-Vidal *et al.* 2021, *A&A*, 646, A52
C. Nanci *et al.* 2022, *A&A*, 663, A129
R. Nesci 2022, The Astronomer's Telegram, 15248
A. Plavin *et al.* 2020, *ApJ*, 894, 101
A. Plavin *et al.* 2021, *ApJ*, 908, 157
A. Plavin *et al.* 2022, The Astronomer's Telegram, 15247
A. C. S. Readhead 1994, *ApJ*, 426, 51
E. Ros *et al.* 2020, *A&A*, 633, L1
M. C. Shepherd 1997, Astronomical Data Analysis Software and Systems VI, 125, 77
J. Stevens *et al.* 2012, Fermi & Jansky Proceedings - eConf C111110, arXiv/1205.2403
F. Tavecchio *et al.* 2015, MNRAS, 451, 1502

The Multimessenger Chakra of Blazar Jets
Proceedings IAU Symposium No. 375, 2023
I. Liodakis, M. F. Aller, H. Krawczynski, A. Lähteenmäki & T. J. Pearson, eds.
doi:10.1017/S1743921323000789

Redshift determination of blazars for the Cherenkov Telescope Array

E. Kasai[1], P. Goldoni[2], S. Pita[3], C. Boisson[4], M. Backes[1,5], G. Cotter[6], F. D'Ammando[7] and B. van Soelen[8]

[1]Department of Physics, Chemistry & Material Science, University of Namibia, Private Bag 13301, Windhoek, Namibia

[2]Université Paris Cité, CNRS, CEA, Astroparticule et Cosmologie, F-75013 Paris, France

[3]Université Paris Cité, CNRS, Astroparticule et Cosmologie, F-75013 Paris, France

[4]LUTH, Observatoire de Paris, PSL Research University, CNRS, Université Paris Cité, Meudon, France

[5]Centre for Space Research, North-West University, Potchefstroom 2520, South Africa

[6]University of Oxford, Oxford Astrophysics, Denys Wilkinson Building, Keble Road, Oxford, OX1 3RH, United Kingdom

[7]INAF - Istituto di Radioastronomia, Via Gobetti 101, I-40129 Bologna, Italy

[8]Department of Physics, University of the Free State, Bloemfontein 9300, South Africa

Abstract. Blazars are the most numerous type of observed high-energy gamma-ray emitters. However, their emission mechanisms and population properties are still not well-understood. Crucial to this understanding are their cosmological redshifts, which are often not easy to obtain. This presents a great challenge to the next-generation ground-based observatory for very-high-energy gamma rays, the Cherenkov Telescope Array (CTA), which aims to detect a large number of distant blazars to study their intrinsic emission properties and to place tight constraints on the extragalactic background light density, amongst others. The successful investigation of these subjects needs a precise redshift determination. Motivated by these challenges, the CTA redshift task force initiated more than 3 years ago a spectroscopic observing program using some of the largest optical and infrared telescopes to measure the redshifts of a large fraction of blazars that are likely to be detected with CTA. In this proceedings, we give an overview of the CTA redshift task force, discuss some of the difficulties associated with measuring the redshifts of blazars and present our sample selection and observing strategies. We end the proceedings with reporting selected results from the program, the on-going collaborative efforts and our plans for the future.

Keywords. galaxies: active, BL Lacertae objects: general, galaxies: distances and redshifts, gamma-rays: galaxies

1. Introduction

Blazars are a type of active galactic nuclei (AGNs) characterized by highly powered relativistic jets that are directed nearly along the line of sight of an observer. They are divided into flat spectrum radio quasars (FSRQs) and BL Lacertae objects (BL Lacs). In the optical regime, FSRQs are characterized by strong and broad emission lines whereas BL Lacs are continuum-dominated with their spectral features having equivalent widths (EWs) of less than 5Å (Urry & Padovani 1995). The very-high-energy emission from

Table 1. Instrument name, observatory name, diameter of primary mirror, wavelength coverage, spectral resolution and combined total number of sources for which spectra have been published in the first two papers by the task force and are to be published in the upcoming third paper.

Instrument name	Observatory name	Mirror size (m)	Wavelength coverage (Å)	Spectral resolution ($\lambda/\Delta\lambda$)	Total sources
Keck/ESI	W. M. Keck	10	3900 – 10000	~ 10000	14
SALT/RSS	Sutherland	11	4500 – 7500	~ 1000	15
NTT/EFOSC2	La Silla	3.5	3860 – 8070	~ 500	18
Lick/KAST	Lick	3	~ 3500 – 5600 ~ 5400 – 8000	~ 1000 ~ 1500	37
VLT/FORS2	Paranal	8.2	3300 – 11000	260 – 1600	14

blazars enables the study of the extragalactic background light density and investigations of the intergalactic magnetic field, Lorentz invariance violation as well as theories of the existence of axion-like particles (Cherenkov Telescope Array Consortium *et al.* 2019, and references therein). The successful investigation of each of these topics heavily relies on blazar redshift measurements, an exercise that is quite difficult to conduct. In fact, excluding FSRQs and blazar candidates of unknown type, 1897 of 2951 *Fermi*-LAT detected blazars have no redshifts. It is thus of utmost importance that efforts are made to measure as many redshifts as possible for general blazar science with CTA. For this reason, a task force was formed within the CTA Consortium more than three years ago, to measure the redshifts of many blazars across cosmic time.

This proceedings is organized as follows: we give an overview of the CTA redshift task force in Section 2, followed by a discussion of the difficulties we face in measuring the redshifts of blazars in Section 3. We describe our sample selection process and observing strategy in Section 4 and present selected results in Section 5. In Section 6, we provide a short summary and briefly discuss our plans for the future.

2. The CTA redshift task force and its objectives

The CTA redshift task force (hereafter the task force) is a CTA working group primarily created with the goal to measure the redshifts of blazars using some of the largest optical and infrared telescopes to support the CTA Key Science Project on AGNs of blazar type (Cherenkov Telescope Array Consortium *et al.* 2019). Table 1 provides a list of some of the spectroscopic instruments and telescopes used by the task force. In the last column of the table are the combined total numbers of sources observed by each telescope, published in the first two papers by the task force (Goldoni *et al.* 2021; Kasai *et al.* 2023) and are *to be published in the upcoming third paper.*

Further details of the instruments in Table 1 are accessible via the websites of the observatories that are home to the telescopes on which they are mounted.

3. Blazar redshift measurement difficulties

The challenging aspect in measuring blazar redshifts stems from the overwhelming non-thermal emission of the jet that, for the most part, outshines the host galaxy thermal emission. While this is not the case with FSRQs that are characterized by strong emission lines, it is so in the case of BL Lacs, whose spectra are characterised by weak host galaxy spectral features that are often quite difficult to detect, even with the best optical or infrared telescopes. The probability of a successful redshift measurement is thus difficult to estimate a-priori.

Because of such challenges, the results of this type of observations are often difficult to interpret. This leads to conflicting conclusions among the various groups of researchers involved in this type of work. Compiling a detailed report on each observed source, including classification details, evidence of extension and previous observational results, is therefore a crucial undertaking. In our attempt to get around such difficulties and still be able to measure the redshift of a given blazar or obtain a lower limit of it, we apply several strategies. These include (1) obtaining high S/N (≥ 100) spectra, (2) searching for absorption systems along the line of sight, (3) detection of host galaxies in deep imaging observations and (4) conducting spectroscopic observations during periods of low optical blazar activity.

4. Sample selection and observing strategy

Our sources were selected from the third *Fermi*-LAT catalog of high energy sources – abbreviated 3FHL (Ajello *et al.* 2017) – at the onset of the project more than three years ago. The catalogue had 1040 BL Lac candidates and BCUs (blazar candidates of unknown type) without redshifts. The selection process was performed with the help of Monte Carlo (MC) simulations using the Gammapy software† to estimate the minimum CTA observation time for each source that yields a 5σ detection. Exact details of the process are presented in Goldoni *et al.* (2021). The sample selection process resulted in the selection of 165 sources that CTA is able to detect in under 30 hours at 5σ.

Our observing strategy is centred on searching for host galaxy stellar absorption and emission features in the spectra. To achieve EW sensitivities of less than 5Å for both absorption and emission features, we require that each spectrum has (1) a spectral resolution $\lambda/\Delta\lambda$ of 1000 or higher and (2) an average S/N ratio of 100 or higher per pixel. For each source, we look for previous spectroscopic results and evidence of extension in the literature, e.g. the 2MASX catalogue (Jarrett *et al.* 2000). This leads to the classification of sources as high-priority or low-priority. Figure 1 shows a flow-diagram summarizing our sample selection process and observing strategy.

5. Selected results

We present selected results from the two papers published by the task force and preliminary VLT/FORS2 spectra of the blazar PKS0903-57 with an uncertain redshift. Goldoni *et al.* (2021) presented spectra for 19 sources, eleven of which had their redshifts determined successfully. Kasai *et al.* (2023) presented spectra for 33 sources (including 13 spectra for eight sources that were of lower quality and featureless). Fourteen of the sources had their redshifts determined successfully. The left side of Figure 2 shows observational results for one source published in the second paper and for which a redshift (given in the subfigure title) was measured. The right side of the figure shows three preliminary VLT/FORS2 spectra of the blazar PKS 0903-57.

At the time of this proceedings, the analysis of the three spectra to determine the significance level of the detected emission feature – marked with the vertical dashed line in the figure – has not been finalized. Further observations are needed to determine the redshift of PKS 0903-57.

6. Summary and future work

To summarize, the task force is a CTA working group tasked with measuring redshifts of blazars that are likely to be detected by CTA and seen as an important support arm for the CTA Key Science Project on AGNs of blazar type. We continue spectroscopic and photometric observations of our sources, and the publication of our results. Our first imaging paper, third spectroscopy paper and a public web database of our observations are currently in preparation.

† `https://gammapy.org`

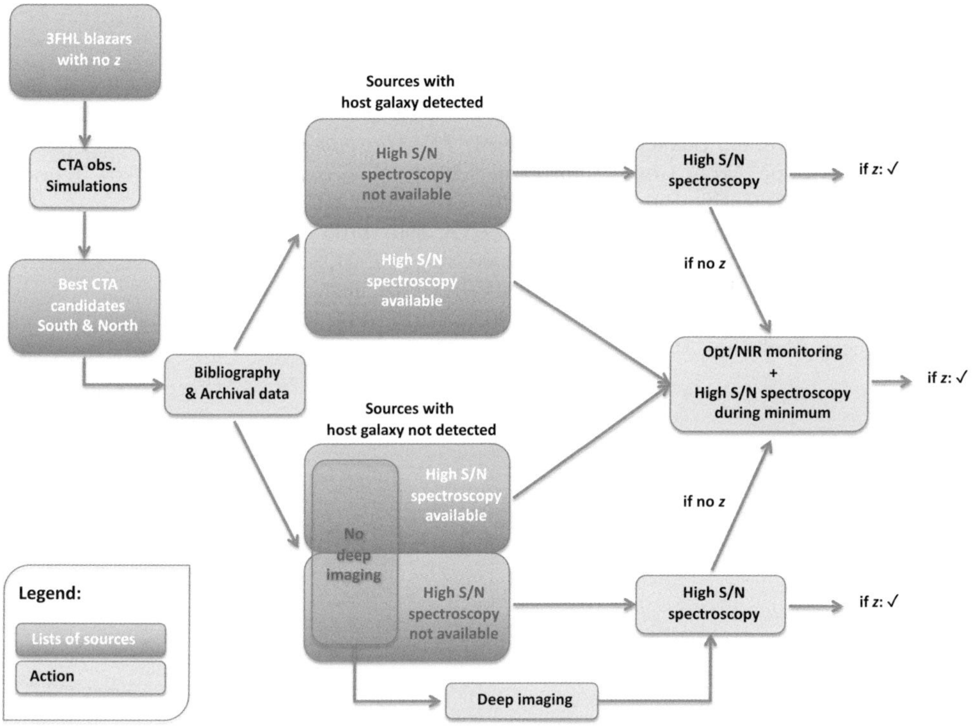

Figure 1. A flow-diagram summarizing the CTA redshift task force source sample selection process and observing strategy.

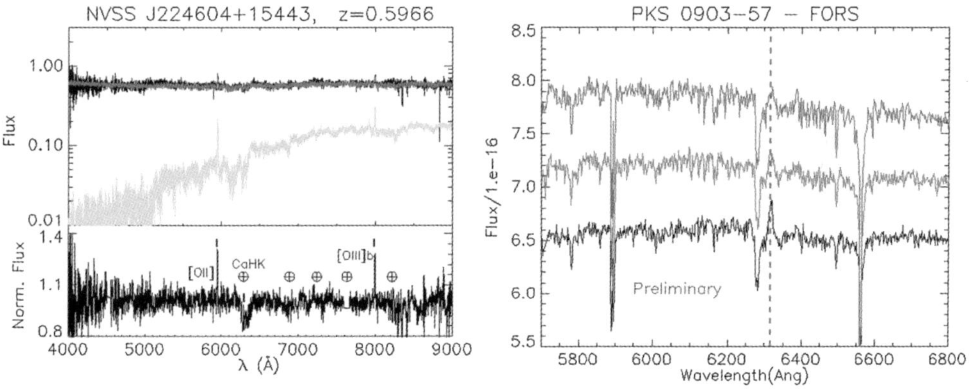

Figure 2. *Left*: Observational results for the source NVSS 224604+15443 published in the second paper (Figure 5, Optical Spectroscopy of Blazars for the Cherenkov Telescope Array – II, Kasai *et al.* 2023, MNRAS, 518). In the *upper* panel are the flux-calibrated and telluric-corrected spectrum (black) and jet+galaxy best-fitting model (red). The elliptical galaxy component of the model is shown in green. The *lower* panel contains the normalised spectrum. The detected spectral features are labelled accordingly. The symbol ⊕ represents atmospheric telluric absorption features. *Right*: Three preliminary VLT/FORS2 spectra of the blazar PKS 0903-57, with an uncertain redshift. An emission feature, marked by the dashed red line, is visible. The strong absorption features around 5890 Å and 6560 Å are, respectively, due to the Milky Way ISM and to a nearby star which lies at 0.67 arcseconds and whose spectrum is superposed onto that of the blazar.

References

Ajello, M., Atwood, W. B., Baldini, L., *et al.* 2017, *ApJS*, 232, 18
Burgh, E., Nordsieck, K. H., Kobulnicky, H. A., *et al.* 2003, SPIE Conference Series, 4841, 1463
Buzzoni, B., Delabre, B., Dekker, H., *et al.* 1984, *The Messenger*, 38, 9
Cherenkov Telescope Array Consortium *et al.* 2019, Science with the Cherenkov Telescope Array, World Scientific Publishing Co. Pte. Ltd., ISBN #9789813270091
Goldoni, P., Pita, S., Boisson, C., *et al.* 2021, *A&A*, 650, A106
Jarrett, T. H., Chester, T., Cutri, R., *et al.* 2000, *AJ*, 119, 2498
Kasai, E., Goldoni, P., Pita, S., *et al.* 2023, *MNRAS*, 518, 2675
Sheinis, A. I., Bolte, M., Epps, H. W., *et al.* 2002, *PASP*, 114, 851
Urry, C. Megan, Padovani, P. 1995, *PASP*, 107, 803

Author index